# The Pushing Field Theory

## Third Edition

Clement Roberge

Could there be more to discover with a pushing gravity model? As an amateur physicist, I have researched this topic and found new information that suggests we should consider this alternative to our current theories. Following a trail of evidence, I have assembled a new theory that, surprisingly, duplicates fundamental principles of physics.

What follows is a fascinating journey into an alternative model of physics with amazing conclusions.

CLEMENT ROBERGE

# CONTENTS

# ACKNOWLEDGEMENTS

I would like to say thanks to the following people who have helped me in writing this book. To my fiancée, Julie Martel, for supporting my efforts in this project. To my kids, Cassandra and Cory Roberge, for their support. To Louis Rancourt for his research and experiments. And to Robert Cahill for editing the book.

# INTRODUCTION

For centuries, physicists have been theorising on the origins of gravity. Isaac Newton was the first to formulate an equation for gravity, and Albert Einstein made further progress with his theory of general relativity. Along the way, however, there have been interesting attempts to offer alternative theories for gravity. One such alternate explanation was proposed by George Louis Le Sage. He was the first to publish a paper on a type of pushing gravity that could be the result of a repulsive force acting on matter. The theory, although interesting as a concept, was never adopted by physicists, and it was left to a few unorthodox professors as an intriguing topic for students to contemplate. An experiment conducted by Louis Rancourt that involved using a laser to lessen the weight of an object suggests that the fundamental mechanism of gravity could be repulsive. After an investigation into the subject, I believe Rancourt's experiment supports a plausible alternative model for gravity. In this book, I will explore the concept of a repulsive gravity and formulate equations. This exploratory journey with a pushing gravity concept has yielded some interesting results. One of the consequences of a repulsive gravity model would be the mechanism for inertia. Following the logical steps in this thought experiment revealed an entirely new universe to explore.

When contemplating the effects of a repulsive gravity in other areas of physics, it became apparent that this idea had more far-reaching implications then just gravity. This investigation into this new concept evolved into a unifying theory. Gravity, inertia, strong nuclear force, gyroscopic effect, precession, propagation of the photon and even the flow of time itself all are components of this theory. I believe a repulsive gravity offers benefits and should be explored further. I encourage others to consider a pushing gravity. This alternative theory may be the most fascinating journey you will take in a new world of physics that has yet to be explored.

# HISTORY OF A REPULSIVE GRAVITY

This story starts in 1687, when Issac Newton published Principia. He theorized the mathematical equation of the law of gravity. Newton was the first to formulate an equation for gravity, but there have been numerous other attempts made by physicists to explain this phenomenon. In 1748, George Louis Le Sage published his theory of gravitation, where he proposed that gravitational forces were the result of small particles impacting matter from all directions. These particles, which he called "ultra-mundane corpuscles", were in a constant state of movement from all directions in space. These particles, as they came in contact with matter, would deliver a force, acting like pressure on the surface of matter.

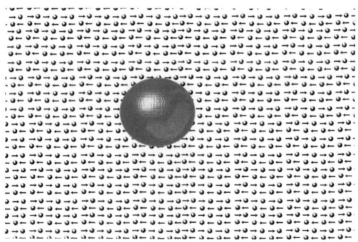

Figure 1

*(The figure above illustrates particles moving in the same plane from either direction would produce opposing forces on either side of the object, represented by the sphere)*

These particles would deliver an impact force to the object. This force acting on the object would be even from all directions. A single object in space would be subject to a force surrounding it, as represented in the figure below.

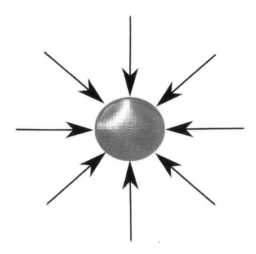

Figure 2

*(The figure above illustrates all forces, represented by vectors, acting on the object through these small particles through impacts from all directions)*

If a second object was in close proximity to the first object, both objects would shield each other from this bombardment of particles. This would result in both objects experiencing an uneven force from the ultra-mundane corpuscles. This force delivered by the particles would be weaker between both objects. The result of this imbalance in forces acting on the objects would resemble an attractive gravity force between both objects. The intensity of this pushing force from the particles would intensify as an inverse square law to the distance, mimicking the effect of gravity. This idea of a pushing gravity was a mirror image of Newton's gravitational model, but the force is not generated from within matter. The end result is a mechanism that is indistinguishable from a attractive gravity model. This model for gravity would have the exact same effect as gravitational attraction.

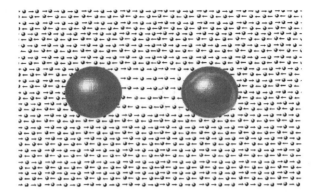

Figure 3

*(The figure above illustrates how two objects would block ultra-mundane corpuscles from moving in between each other.)*

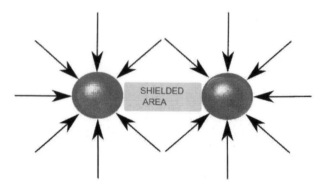

Figure 4

*(The figure above represent the forces caused by the ultra-mundane corpuscles on both objects. The shaded area represents the region of less force on the objects)*

LeSage's theory on gravity was, for the first time, a published paper explaining how gravity could be generated using a repulsive force instead of an attractive force. There were a few problems with this model, however, that could not be reconciled, one of which is the excess heat generated by the impact of the particles on the object. The extra heat was never measured, and the theory was later abandoned when no solution to the problem was found.

It is believed that the mathematical equation and the idea for a pushing gravity proposed by Le Sage was originally put forth by Nicolas Fatio de Duillier, a mathematician who worked on the idea in the late 1600s but never published his work. Nicolas Fatio de Duillier was also a good friend of Isaac Newton. It has been postulated that the idea of a pushing gravity may have originated from Newton himself when we was contemplating the idea of gravity. This must have been a topic of conversation between both mathematicians. Fatio, intrigued by this idea, would have explored this hypothesis mathematically. He would have talked about it to some of his close friends, one of whom was George's father. George new of Fatio's work and, upon his death, bought Fatio's research notes but never credited Fatio for his work. It is interesting to see how Newton was potentially involved from the start with both gravity theories.

Another idea that originated around that time was that of the Aether. Isaac Newton first hinted in 1718 at the possible existence of the Aether, a substance that filled space with small particles. This is another reason to suspect that Newton must have considered a repulsive gravity. The assumption of the existence of the Aether is a logical step when considering a repulsive gravity. This idea of this Aether did not stop there, as throughout the 18th and 19th centuries, physicists continued to use the Aether to explain certain phenomena. From light propagation to gravity transmission, it became the source of many postulations. In the early 1900s, the Luminiferous Aether theory was one of the last attempts by physicists to use the Aether to explain the propagation of light. Many attempts were made to prove its existence. In the end, Michelson-Morley's failed experimental attempt to prove the existence of the Aether set the stage for Albert Einstein's theory of relativity. The Aether theory and LeSage's gravity theory have faded into obscurity, as they were not ideal theories to represent the mechanisms at work in the universe. These theories, however, could be stepping stones in the right direction. In 1865, another great scientist, James Clerk Maxwell, published "A Dynamical Theory of the Electromagnetic Field". Many of our technological advancements can be attributed to his work. There have been attempts to reproduce his success by mimicking what he did in respect to the electromagnetic field with gravity. These works did not produce significant advancements, but they do bring about an interesting question: What if a similar attempt was done with a repulsive gravity? As far as I know, this has never been attempted. Let us proceed with this idea and see what we find.

# A DYNAMICAL THEORY OF A PUSHING GRAVITY

Before getting into equations similar to Maxwell's, we have to revise some of the properties of a repulsive gravity model. Fundamentally, LeSage's gravity force is generated from space. It is throughout the universe in somewhat equal strength, pushing in all direction, and down on the earth. This force is generated by the emptiness of space. Every point in the universe is creating this force that moves outward from the point of origin.

This force appears to be repulsive. The effect of gravity results from a process similar to Le Sage's theory of gravitation. To differentiate from past failed attempts and eliminate confusion by using the word "gravity" to describe this effect, a new name will be given to this new idea. This name should describe the process that is happening. A pushing gravity is what is produced, so the field that produces this effect will be called a "pushing field". A pushing force is created everywhere in equal quantity from this pushing field. This can be tricky to imagine. An easy way I have found to visualise the process of a pushing field is by imagining ripples of waves traveling on the surface of water. These waves are generated at a single point and emanate outwards like ripples from a drop of water. You can imagine rain on a calm lake generating ripples everywhere, traveling in all directions.

Figure 5

*(The figure above illustrate a ripple from a raindrops that represents a force generated from a single point)*

These ripples in the pushing field are invisible to the naked eye, but if we could see them, we would see a fluctuating field. The waves propagate out from their point of creation to encounter other waves from all directions. The field would appear to be a series of crests and troughs. To see the effect of that field, we must pick a point that is relevant and visualise a force coming from that one point as waves move outwards. As the waves encounter matter, a force is delivered. We can represent this force with vectors.

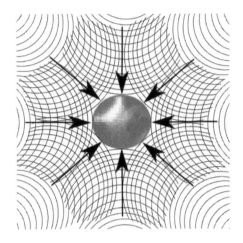

Figure 6

*(Vectors representing the force from the pushing field being applied to matter, represented by the sphere)*

For example, if we use the earth and choose a point not to far from it, we can visualise waves coming from that one point toward the earth. Any object positioned between the earth and that point would experience a force from the waves accelerating it towards the earth, similar to a beach ball being pushed by the waves of an ocean towards the beach. These waves are massless but carry energy, so they will travel at the speed of light through space. We can now attempt to formulate this new model for the pushing field.

The pushing field can be represented geometrically with a positive divergence from a single point. An example of this vector function is shown in Figure 7; it comes from a single point and flows out. Field lines can be drawn that demonstrate the direction of the force.

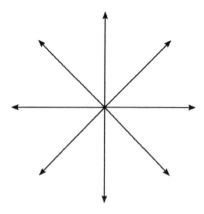

Figure 7

*(Field lines diverging from a single point)*

As the line moves outward spherically, the density of the lines from the center point decreases. There is not just one single point diverging in space. Space is made up of an infinite number of points of these field lines with a flux density. Every point in space will have an infinite number of field lines to produce the pushing flux of the pushing field through an area.

## Pushing force resulting from the pushing field

Each point in the field fluctuates from a temporary change in energy. The fluctuation will produce oscillations in the field lines. The oscillation starts from the point of origin and moves radially outward. This resembles a plucking of the single point of origin, much like a pick hitting a guitar string. The oscillation starts at the point of origin and moves out from that point down the field line at the speed of light. All the field lines diverge from that single point, oscillating at the same time as they travel radially outward from the point of origin. This process resembles a ripple on the surface of water. The vector function of the pushing field travels radially outward from the point of origin. The field strength drops at a rate of the inverse square of the distance from where it is emitted. The vectors get smaller the further the distance from the point of origin.

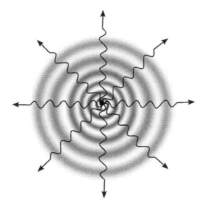

Figure 8

*(Field line oscillation diverging outwards at the speed of light, resembling ripples)*

The waves will produce constructive and destructive interference as they propagate through space. The behavior of these waves is similar to waves on the surface of water. Two point sources in space will produce propagating waves and the resulting amplitude will be equal to the sum of each individual amplitude. The combination of the linear expansion of multiple points traveling in the same direction will produce plane waves. The direction of the waves is the direction of the energy flow, resulting in a propagating vector wave. When the propagation of the wave is blocked, the energy transfer from the wave will result in the form of a force. Matter will block the propagating wave. When the waves in the field lines encounter the nucleus of an atom, they are prevented from progressing any further. The energy in the wave is released by a force exerted on the nucleus of the atom. The nucleus is only a small portion of the atom, and the remainder of the waves passes through the atom with no interference. The force on the nucleus of an atom is proportional to the number of field line waves being blocked by the nucleus. Each atom contributes to the force received from the pushing field by blocking the waves. Therefore, the force on the object is a combination of each individual force on the nucleus blocking the waves from the pushing field.

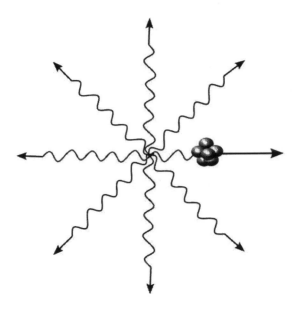

Figure 9

*(Representation of a particles blocking a single field line oscillations from propagating)*

The nucleus of each individual atom will determine how much force will be received from the pushing field. Once the waves have been blocked by the nucleus, the field will start once again to oscillate to form new waves travelling away from the nucleus. The strength of the force from the newly created waves will accumulate in strength with increased distance from the nucleus. The force potential from the newly created wave increases at a rate according to the inverse square law as a function of distance away from the nucleus. The interaction between the matter and the pushing field can be calculated using Equation (1). This equation is very similar to Newton's formula F=ma. The force needed to accelerate an object is the force the pushing field exerts on the object from the opposing side (This will be explored in detail later in this book). The blocking of the opposing waves from the field in an indirect way causes the "mass of the object"; therefore, the mass has to be divided in half to represent opposing sides. The acceleration in the formula is simply replaced by the speed of light, as that is the velocity of the propagating wave. The wave starts from a zero velocity and increases to the speed of light; this is in a way its acceleration. To complete the formula, this interaction between matter and the field is continuous and happens as an interval of time. This equation gives the strength of the force from the pushing field acting on the same plane from one of the opposing side of the mass. It is the force delivered by the waves on to matter.

$$F_{Pf} = \frac{c\,m}{2(t)}$$

(1)

$F_{Pf}$ = Force from the pushing field (Newtons)

c = Speed of light (m/s)

m = Mass (kg)

t = Time interval of 1 second (s)

As Equation (1) demonstrates, the amount of force exerted on matter is enormous. This creates a very strong force acting on the nucleus of the atom, but when encompassing the entire atom, the blocking effect is minimal. The force acting from all sides will produce a negative divergence force acting on matter. A similar comparison can be drawn from the interaction between matter and the pushing field and the interaction between a charge and the electric field.

From this interaction between the pushing field and matter, all kinds of phenomena are produced. These can be classified into two groups: static and dynamic phenomena. A static phenomenon is caused by the presence of matter within the field. I believe gravity, and the strong nuclear force are produced by matter blocking the pushing field waves. The second group of phenomena result from a change in position with respect to the pushing waves as they act on matter. In this group, we find the effect of inertia. This group consists of acceleration and deceleration force, centripetal force, gyroscopic effect, and precession. I will explore in detail some of these phenomena in later chapters of this book. But first, let us explore gravity.

# GRAVITY

The gravitational force experience by objects is a leftover from two opposing pushing forces acting on them. The forces from the pushing field waves will decrease in strength as it passes through an object. As a result, the force from the opposing waves that has not yet pass through the object will be stronger. This leaves all objects with a net force acting toward them. This process would produce a force that resembles an attractive force we know of as gravity. The effect is caused by the pushing waves passing through matter and being reduced by it.

The waves from the pushing field propagate at the speed of light in every direction. The presence of matter in the field causes the waves to be blocked, stopping it from propagating. A differential force will develop caused by the opposing waves travelling in the opposite direction. The difference between the two opposing waves is the force of gravity. Matter is actually being pushed together because it reduces the force passing through, and the oncoming force is greater.

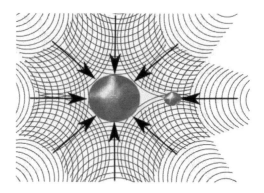

Figure 10

*(Waves are blocked from propagating by matter. Opposing waves push matter together to create a gravity effect.)*

This differential force exists all around any object. Any two objects in close proximity to each other will cause a differential force that will move both objects toward each other.

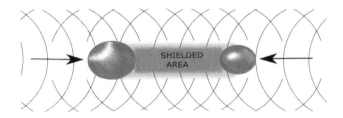

Figure 11

*(Pushing waves are blocked by objects, creating a shielded area. The opposing waves will push both objects together.)*

This becomes the fundamental mechanism that produces the effect of gravity, as opposing vectors now have a differential potential. Any two object in space will produce a differential force on each other. The blocking of the waves from one object will cause in a differential force across a second object. In addition, that second object will cause a differential force across the first object. This produces a differential vector force on both object. The differential force acting on both object will cause them to move toward each other. The differential in those forces will result in a force that will "push" both objects together.

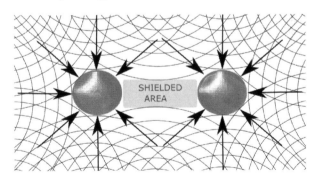

Figure 12

*(Differential force on objects by blocking the waves)*

This mechanism that produces gravity could potentially be producing other forces. A nucleus of an atom will block most, if not all of the oscillations of the field lines. This would produce a very strong pushing force on the nucleus. This could possibly be part of the mechanism that keeps the nucleus together and is referred to as the "strong nuclear force" in physics. The only other constituent of an atom is the electrons. It would not produce significant blockage of the pushing waves passing through an atom. Therefore, the nucleus is the only part of an atom that would produce a force as the pushing waves passes through an atom. This force decreases rapidly when moving away from the nucleus and is substantially lower at the edge of the atom. When considering the differential force caused by the nucleus to the size of an atom, it is the same as comparing the strong nuclear force to the force of gravity. The force of gravity is substantially lower in strength then the strong nuclear force because atoms are not blocking any additional waves. It is the area that is considered that is different. It is possible that the force produced by the nucleus can be the strong nuclear force and also the force of gravity. Gravity is the compounding effect of the residual forces from all nucleuses within an object. This forces extend outward and decrease in strength to the inverse square law, producing gravity. To find the force of gravity using the pushing field equations, Equation (1) can be modified to include the inverse square of the distance from the mass. A new constant is introduced for the equation to work. This new constant is call the pushing field gravity constant and is simply the speed of light divided by the Newton gravitational constant.

$$\varsigma = \frac{c}{2G}$$

(2)

c = Speed of light 299,792,458 m/s

$\varsigma$ = Pushing field gravity constant
($2.24597 \times 10{-}18$ kg·s·m$^{-2}$)

G=Gravity constant $6.674 \times 10{-}11$
m3·kg$-1$·s$-2$

The gravitational force potential from the pushing field can be calculated using equation (3).

$$F(g) = \frac{cm}{2\varsigma r^2}$$

(3)

F(g)= Gravitational force resulting from the differential forces of the pushing field (Newtons)

c = Speed of light (m/s)

m = Mass (Kg)

ç = Pushing field gravity constant $2.24597 \times 10{-18}$ kg*s/m$^2$

r = Distance (m)

To calculate the force of gravity between two masses, Equation (3) was modified to represent $m_1$ as the first mass and $m_2$ as the second mass. The distance between both masses is **r** in meters. The effect will decrease as the distance between two objects increases. The equation (4) is the force of gravity **(Fg)**, which is the force from the pushing field on one mass **(m1)** multiplied by the second mass **(m2)**, over the distance that separates them in meters **(r)** and the pushing field constant **(2.25*10$^{18}$ kg/m$^2$)** .

$$F(g) = \frac{c\, m_1 m_2}{2\, \varsigma\, r^2}$$

(4)

The equation is the same as Newton's gravity formula with a few extra terms added. The addition of these terms does not change the outcome, but it does demonstrate that the pushing field force from equation(1) is fundamental for producing the force of gravity.

That being said, there is a problem with Newton's gravity formula: It does not account for Mercury's precession. First noted in 1859, Mercury's orbit disagrees with Newtonian theoretical predictions. A closer approximation of the motion of the planet was given by Einstein's general relativity theory. Why am I reconsidering this formula for gravity? Because there is more to consider than just a force of gravity as demonstrated by Einstein's theory.

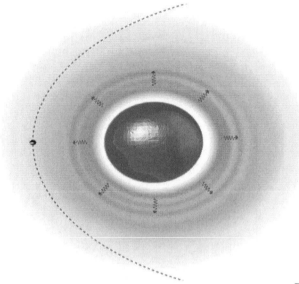

Figure 13

*(Mercury's orbit near the sun changes trajectory every time it passes by)*

## Anomalous Perihelion Advance of Planets

The anomalous perihelion of Mercury can be explained differently and yet similarly to Einstein's theory of general relativity. The idea will mimic general relativity in some ways. Einstein proposed that space and time are curved near matter. The theory proposes that objects in the vicinity of a planet will experience more "time". Their clocks will run slower near matter compared to clocks further away. In a way, Mercury takes more time than is expected as it passes by the sun. Time is a factor that cannot be dismissed.

There is another component of the pushing field that exists and can produce the relativistic effect theorised by Einstein. But before explaining this, we first have to continue exploring the pushing field and how it may play a role in other phenomena.

# INERTIA

All objects in the universe are affected by the waves from the pushing field. Each wave is opposed by another wave. When the waves encounter an object, they deliver the force from the waves. If the force is of equal value, then the forces are nullified and have no net effect on the object. In other words, the object is held tightly in its place by the opposing forces from the pushing field. To make a change in the object's motion or velocity, a force is required to overcome the force from the field that is holding the object tightly; this is the mechanism of inertia. This effect is due to the "synchronization" of the pushing waves on objects.

To understand the concept further, visualize a stationary object located exactly halfway between two points in space. Waves are emitted from both points at the same time, and these waves travel toward the object. The waves should arrive at the object at the same time. If the waves are carrying a force, then the forces from the waves are equal but moving in opposite directions. The force from the waves would cancel each other out. Now, let us add motion. If the object were to move towards one of the points before the waves contacted the object, then the object would contact one of the waves before the other opposing wave. The object would feel a momentary imbalance in the forces resulting from the asynchronous waves. It would first experience the wave in the direction in which it was moving; then, a moment later, it would experience the force from the other direction.

Figure 14

*(The waves moving toward the object are no longer synchronized, as the object is moving away from one wave and toward the other. This change in position causes an imbalance in the force from the pushing waves.)*

Let us add more complexity. The two points are emitting a series of waves. With the continuous waves coming from the two points, the object will only feel the first waves it encounters. The wave in the opposite direction will be opposed by the second wave from the direction of travel. With continuous waves, the only wave not being opposed is the first wave the object experienced because of the change in motion. This leaves the object with a force working against it in the direction of movement. This is the force it has to oppose to move in that direction. It is also the force it will feel when the movement occurs. This is a phenomenon referred to as inertia, and it is experienced when there is a change in velocity or a change in motion.

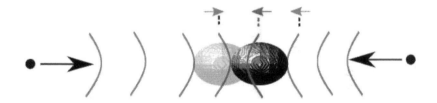

Figure 15

*(Continuous waves result in the object feeling only a force in the direction of travel when accelerating in that direction.)*

To further explore this concept, if the object were to move twice the distance it previously did, then it would receive twice the amount of force. The amount of distance the object traveled in a set period of time would be proportional to the amount of force it would receive from the pushing

19

field. In other words, if the object moved twice as fast as it previously did in one direction, then it would cover twice as much distance in the same amount of time. The object would receive twice as many waves than it otherwise would have in that direction. This would require the object to double the force required to achieve this velocity. The greater the distance an object has to travel in a set amount of time, the greater the force it will require to reach the necessary velocity.

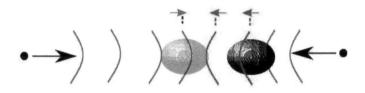

Figure 16

*(An increase in acceleration results in a proportional increase in the opposition from the field.)*

This concept still applies if the object already has velocity. In this scenario, visualize the same two points, but the object is not in the middle of the two points but offset to one side. The object is moving away from one point and toward the other point. The two points are emitting continuous waves. When the waves reach the object, the object has travelled from its initial starting position to the midpoint. At that location, both waves are equal in strength and opposing each other. The object does not feel anything at that location from those two points. The forces from the opposing waves are cancelled out. For an object that already has velocity, the wave can still be synchronized. Each wave has an opposing wave to cancel it out. This keeps the object in the state it is in. If the object was to accelerate in respect to the same scenario previously discussed, it would still encounter the oncoming wave first. This would leave the object with a differential force as it accelerates. It is only in a change of motion or velocity that the object would feel the force from the imbalance in the pushing waves.

## CONSTANT VELOCITY

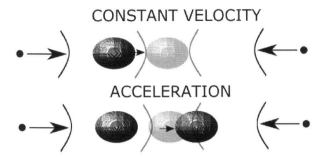

## ACCELERATION

Figure 17

*(Even if an object has motion, the same concept of inertia resulting from opposing waves will produce the same effect. An object with motion is not exactly in-between both points but offset to one side. When the waves reach the object, the object has moved to the center position and is synchronized. If the object increases in speed, the waves will no longer be synchronized.)*

It can be overwhelming to consider an object's motion while waves are coming from all directions. It can be mind boggling to visualize each wave contacting the object at different times and the effect this has on the object. The trick is to break it down to a very specific scenario. To make it easier, eliminate all aspects that have no consequences and then go through the process step by step. This will make it easier to visualise. The same concept of cause and effect can be applied to all scenarios of inertia. Let us analyze how an object experiences deceleration.

When an object reduces its velocity, this reduction in velocity will change the synchronization of the waves as it contacts the object. Using the same previous scenario, the object is moving between the two points. The object is closer to one of the points and is moving toward the other point. When the object is in constant velocity, the waves from the two points contact the object at the same time at mid-point. No force is felt from those two points. As the object decelerates, the wave from the rear will contact the object first. This leaves the object temporarily with an imbalance in the forces. The direction of the force will be in the direction of travel. The object will have to overcome this force to be able to reduce its velocity. The extra force encountered from the rear is felt as inertia. Let us consider a moving automobile as an example. As a vehicle reduces its velocity, the occupant also receives extra force and has to resist it. This is the force you feel pushing you forward when coming to a stop. The decrease in velocity causes more waves from the rear to come into contact with the vehicle compared to the front. This is the force felt by the occupants, pushing them forward.

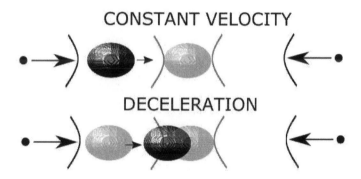

Figure 18

*(The same concept would apply if the object decreased its speed. The object would receive the rear wave before the forward wave as it decreases in velocity and would receive a force as it decelerates.)*

Acceleration and deceleration are relative to the pushing field. If the observer accelerates in a certain direction, he will be gaining speed relative to the field at the moment of acceleration. Because of this, he will experience a force from the field. In a way, as the acceleration occurs, the object is moving away from the rear force and moving forward at an increasing rate. A doppler shift occurs with the waves of the pushing field acting on the object.

The force differential resulting from a doppler shift in the pushing field waves as an object accelerates can be calculated with the equation (5).

(5)

$$F(a) = \frac{cm}{2(t)} \left( \frac{c}{c-(vf-vi)} - \frac{c}{c+(vf-vi)} \right)$$

**F(a)** = force on an object as it accelerates caused by a doppler shift in the pushing field waves (N)

**c** = speed of light (m/s)

**m** = mass (kg)

**t** = time interval of 1s (s)

**vi** = initial velocity (m/s)

**vf** = finial velocity (m/s)

To accelerate an object, the forward force from the waves of the pushing field has to be overcome. This formula will produce a result similar to Newton's F=ma for low velocity. The formula is similar to special relativity in that an increasing amount of force is required to accelerate an object when getting closer to the speed of light. A comparative graph in Figure 19 demonstrates relativistic acceleration versus pushing field acceleration for the same mass. As a result of the doppler shift, mass appears to increase.

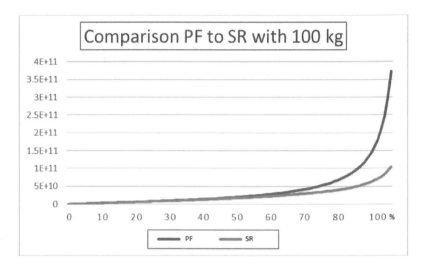

Figure 19

*(Graph comparing relativistic force (SR) compared to the pushing field force (PF) from acceleration over a percentage of the velocity of the speed of light)*

It is interesting to see how a doppler shift can produce the effect of inertia as an object accelerates. This idea does have a wave frequency problem in regard to the amount of force the object will receive at different velocities. When an object increases its velocity, it also increases the amount of waves it will receive from the pushing field. For this model to work, a solution to this problem is provided in the next chapter, and may answer other intriguing questions.

# INFINITE FREQUENCIES IN THE FIELD

Although the proposed mechanism of inertia seems to work, there is a fundamental problem that needs to be resolved. When the object is moving against the waves from the pushing field, those wavelength in the direction of travel will shorten. At the same time, when an object is moving away from the waves of the pushing field, the wavelength the object is moving away from will lengthen. This is equivalent to saying the frequency coming from the forward direction will increase, and the frequency the object is moving away from will decrease. If there was only one frequency in the pushing field, the change in frequencies as a result of the change in the velocity would change the amount of force the object is receiving from the pushing field. The change in forces would work against the object to reduce its velocity. This would defy one of Newton's laws of motion: Every object in a state of uniform motion will remain in that state of motion unless an external force acts on it. A solution to this problem can be found through the existence of an infinite number of frequencies in the pushing field line oscillations. This resolve the one frequency problem caused by velocity change. When changing velocity, there is always another frequency to oppose the original frequencies. When the object increases its velocity, the oscillation is compressed in the forward direction. If there is more than one frequency that exists in oscillations of the pushing field, then there is a matching opposing frequency at a higher frequency. The existence of infinite frequencies in the pushing field would produce a force only when the object transitions from one frequency to another. This shift in opposing frequencies results in a force on the object as it accelerates or decelerates.

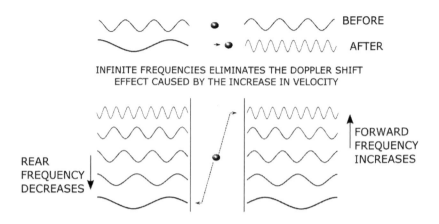

Figure 20

*(By accelerating, an object shifts opposing frequencies. The net force on the object remains the same as the total amount of frequencies acting on the object does not change.)*

The existence of multiple frequencies in the pushing field would explain another phenomenon. The Casimir effect is an attractive force between two plates, and Van der Waals force is an attractive forces between two molecules. These forces are similar to each other but different to the force of gravity and the strong nuclear force. If multiple frequencies existed in the pushing field, then a phenomenon similar to Van der Waals force would result from missing frequencies between two molecules. Longer wavelengths would not "fit" between two molecules, but would be present on either side. The end result would appear to be an attractive force as the plates or molecules are being pushed together. If the distance between the two objects is small to allow the existence of the wavelength, then it cannot be created.

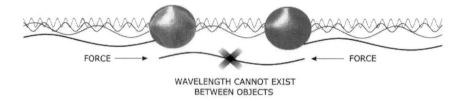

Figure 21

*(Longer wavelengths cannot exist between two objects, and this create another force known as the Casimir effect)*

The wavelengths that cannot be created between the two objects can be created beyond the two objects. This creates a differential force on the two objects; this force is completely different from the force of gravity. A differential force resulting from missing frequencies and not the blocking of the pushing field oscillation.

In my interpretation, I prefer an infinite number of frequencies. An infinite number of frequencies with pushing field lines also eliminates absolute motion. This was a problem with the theory of the Aether. The Aether theory used a substance to fill the universe through which waves propagated through space. One of the problem with the Aether theory was an object with motion could be referenced to the "flow of the Aether", and a measurement could be made to an absolute zero velocity. There has been many attempts to measure absolute motion, but no experiment has ever successful. This problem can be resolved with an infinite frequencies. It does not matter how slow the object is moving, as there is always a frequency that exists within the pushing field lines. In addition, an object may be reducing its velocity or could be accelerating in the opposite direction, there is no way of telling. When compared to a finite number of frequencies, a measurement could be made to the lowest frequency, and a velocity could be obtained. By removing these limits, we are removing any reference to an absolute velocity, allowing no way of telling what velocity an object has. This view would reproduce the relativistic effect theorized by Einstein. The only way of measuring the velocity of an object is in reference to another object. The idea of an infinite number of frequencies in the pushing field provides many solutions to the problems that perplex other theories. This extends to the next chapter with the propagation of electromagnetic waves.

# ELECTROMAGNETIC WAVES & PUSHING FIELD INTERACTION

The interaction between electromagnetic waves and the pushing field can be seen with some of the behaviors of light. The force from the pushing field can move oscillations in the electric field, and at the same time, the those oscillation in the electric field can opposes the pushing field. Evidence for these interaction can be observed in different phenomena. First, light is bent in a gravity field as it passes near a massive object. This is referred to as gravitational lensing.

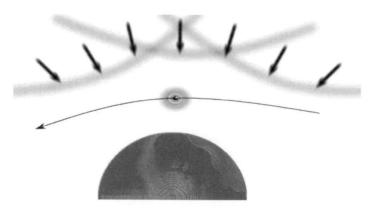

Figure 22

*(Illustration representing gravitational lensing as a photon passes by a massive planet)*

Gravitational lensing has been confirmed by astronomers. As discussed previously, gravity is a phenomenon resulting from the pushing field. Therefore, electromagnetic waves must have an interaction with the pushing field to produce gravitational lensing. An experiment by Louis Rancourt ("Effect of light on gravitation attraction" published in the science journal Physics Essays) also suggests that electromagnetic waves, similar to matter, can block the pushing field and produce a gravity effect.

Another aspect to consider is the velocity of electromagnetic waves. The speed limit of the universe is the speed of light. Nothing can goes faster. Why is that? As for the pushing field theory, the speed of light results from the pushing field's velocity. Physicists do believe that gravity moves at the speed of light. If the pushing field is responsible for gravity, then the speed of light is inherent to the pushing field. The pushing field's velocity is the speed of its actions, and therefore whatever is affected by it. If a particle offers no resistance to one of the opposing forces of the pushing field, it will be pushed by that force at the velocity of the waves, which is the speed of light. A photon would be an example of a particle pushed by the pushing field. The photon is just a passenger riding the waves. Pushed by the forces of the pushing field at the speed of the propagating wave. The waves also guide the photon as it passes through a "gravity field". The uneven forces on either side of the photon will change the path of the photon. This force acting on the photon would change the trajectory as it passes by massive object. How can this interaction happen between an electromagnetic waves and the pushing field waves?

I believe these interactions between both fields can be explained through the oscillations of field lines. When the oscillations are identical (meaning the same frequency), the pushing field lines and the electric field lines are in harmony. There is no opposition and therefore no force is created. When the oscillation are not the same, then a force is created. The force between both field will be in the same direction of the traveling wave. This force will push the photon at the same speed of the propagating wave that is pushing it.

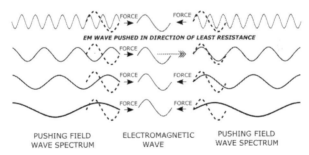

EM WAVE PUSHED IN DIRECTION OF LEAST RESISTANCE

| PUSHING FIELD | ELECTROMAGNETIC | PUSHING FIELD |
| WAVE SPECTRUM | WAVE | WAVE SPECTRUM |

Figure 23

*(Oscillating electric fields are propelled by the pushing field from opposing oscillations. One identical oscillation offers no resistance, while an opposing oscillation pushes on the oscillating electric field to move it in that direction.)*

All frequencies in the pushing field that are not identical to the oscillation of the electromagnetic waves will create a force on the electromagnetic wave. The pushing field frequency that is identical to the frequency of the electromagnetic wave will offer no opposition to the electromagnetic wave. At that frequency, the force from the single opposing pushing waves will move the electromagnetic wave in the direction it is moving in. This mechanism for the propagation of light will function at any velocity. It does not matter at what velocity an object is moving; when a photon is emitted, there is a frequency that exists to carry the frequency of the photon. This is another good argument for the infinite frequencies that exist in the pushing field. All frequencies that exists in the electromagnetic waves spectrum must also exist in the pushing field wave spectrum in order for the pushing field to move them.

In addition, this idea is further supported by the Doppler effect observed in the electromagnetic waves. An electromagnetic wave is emitted at a specific frequency. The emitted frequency of the electromagnetic wave remains "locked in" to the pushing field frequency. Unless the electromagnetic wave is measured at the same velocity and same direction, a shift in the frequency in the electromagnetic wave can be observed. In actuality, the frequency of the electromagnetic wave has remains the same, and it is the frequency used to measure electromagnetic wave that has shifted. The electromagnetic wave propagates at a specific frequency, and is carried by the pushing field at that frequency. Changing velocity also changes the carrying frequency of the pushing field. This also changes the reference used to measure the frequency of the electromagnetic waves.

This is a radical change for current traditional physics. It is currently believed that electromagnetic waves are self-propelled. Identifying the pushing field as the mechanism of movement for the photon is a bold claim, but it does support the theory. This support comes from different areas of physics. One such area is quantum mechanics, which will also be discussed in a later chapter. The other support comes from special relativity and is discussed in the next chapter.

# THE PUSHING FIELD AND SPECIAL RELATIVITY

The failed experiment of Albert A. Michelson and Edward W. Morley marked the end of the Aether and the beginning of a new era. In 1905, Albert Einstein published his theory of special relativity. Einstein's special relativity still holds up today, and it is accepted by the majority of physicists. Einstein utilized a Lorenz transformation equation in his special relativity theory. The Lorenz transformation demonstrated that time and space are linked and may vary depending on the location and velocity of the observer. These observations, referred to as reference frames, dismiss the idea of absolute time. An observer traveling at great speed does not experience time at the same rate compared to a stationary observer. The greater the speed the observer is moving at, the slower that time moves forward.

The pushing field theory shares a similar conclusion in regard to different observations of time made by different observers. This can be concluded in understanding how the pushing field is a mechanism that displaces particles. The field acts like a canvas on which the universe is painted. The pushing field is responsible for the propagation of light, the force of gravity, and the effect of inertia. The field is also responsible for the velocity of light. When a photon is propelled by the force of the pushing field, the photon's oscillation causes it to oppose the pushing waves from the rear and become transparent to the opposite wave from the front. This provides the mechanism that causes the photon to be carried by the pushing waves at the speed of light. The photon follows the path of the pushing waves is taking. The behavior of light is a demonstration of the movement of the waves propagating through the pushing field. The photon follows the movement of the invisible waves from the pushing field. What we understand as some of the properties of light are may really be the properties of the pushing field. The photon is just a passenger being transported by the guiding waves of the pushing field.

31

This mechanism of pushing waves moving in every direction will also produce the effect we know of as the flow of time. As the waves propagate through space, they encounter matter in the form of particles. These particles behave in a particular way as a result of the pushing waves. The force of gravity is created as the particles block the waves from propagating. Inertia results when the particles change their position relative to the opposing waves. Each wave causes something to happen to particles, as each wave produces an event. Each occurrence is sequential, resulting in the flow of events. All events are evenly distributed from one event to another, giving rise to the flow of time. Time is not a separate entity nor its own dimension but just sequences of events that result from the propagation of the field. The experience of time can simply be a sequence of events that happens as the pushing field waves passes by. If one sequence of events is divided into two, then each divergent event will have its own flow of time as it interacts with the mechanism of the pushing field. This is the reason why time is not absolute.

These events occur as a result of waves interacting with particles. The velocity of the particles in respect to the field will dictate how many occurrences happen to the particles. When particles are traveling at a high velocity, they are limited in the number of events that can occur. In other words, time will appear to slow down because there is less interaction that can occur resulting from the pushing waves. The consequence of fewer occurrences is a slower progression of time for particles and objects moving at a high velocity, especially when approaching the speed of light. The relativistic mirror clock thought experiment can be used to demonstrate this concept. Two mirrors are positioned perpendicular to each other, and a photon "bounces" between both mirrors. If the mirrors are stationary, the photon will bounce between both mirrors at the speed of light. Time would flow at a normal rate to an observer looking at the experiment. If the experiment was conducted on a spaceship moving at a high velocity, an external observer would see the photon slow down in-between bounces. But to the observer inside the spaceship, the photon would maintain the same speed. This is called time dilation. Time varies between both observers. This effect can be explained as a result of the photon's increase in distance between both mirrors as the ship moves through space.

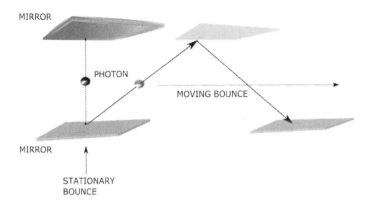

Figure 24

*(Demonstration of the additional distance a photon has to travel between both mirrors as it is moving)*

The photon's velocity is fixed at the speed of light. Since the velocity will not change, then the construct for the flow of time for the photon has to slow down. This is similar to what Einstein theorized with time dilation in his paper on special relativity. This effect has been confirmed experimentally by the Hafele-Keating experiment. Individual atomic clocks measured different time flows as they were subjected to different velocities. The atomic clocks on planes flying through the air ran slower compared to stationary atomic clocks on the ground. This effect is real. The pushing field theory predicts the same "time dilation" experienced by moving objects as described by Einstein's special relativity. The effect of time dilation can be explained by the pushing field theory as a result of an increase in the distance the particles have to travel to accomplish the same event. The pushing field moves the particles a greater distance when the particles are moving at a velocity in a particular direction. This will take an additional amount of time compared to a particle that is at rest doing the same thing. Let us return to the thought experiment of the photon bouncing between opposing mirrors. Consider a photon that is bouncing back and forth between two parallel mirrors on a spaceship. If the two mirrors and the spaceship are stationary, the photon will bounce perpendicularly between both mirrors at the speed of light. Each time the photon bounces off the mirrors, the waves from the pushing field carry the photon to the other mirror. The time between both mirrors can be calculated by the speed of light divided by the distance between them. When both mirrors in the spaceship are moving at a high velocity in one direction, then the waves of the pushing field carrying the photon have to move at an angle to be able to maintain the photon's trajectory perpendicular to the mirrors. As the photon moves from one mirror to another, the distance the photon has to travel increases. This increase in distance can be calculated by the Pythagorean theorem. This follows the same reasoning as the Lorentz transformation.

Therefore, the photon will take longer to bounce between the two mirrors. The photon still moves at the speed of light, but it has to travel a longer distance to get to its destination. Its overall time of travel between both mirrors has increased. To an outside observer, the photon's speed appears to have decreased in velocity, but it is the flow of time that has slowed down.

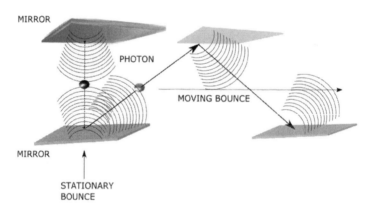

Figure 25

*(The photon bouncing between both mirrors will appear to have slowed down to an external observer as it moves in a particular direction)*

If an observer is present in the spaceship with the two mirrors, he will see the photon moving at the speed of light. The observer is also subject to the extra distance the pushing field has to propagate as the ship moves through space. Each point in space produces a wave rippling outwards. This occurs first, and all other phenomena occur afterwards. Each wave from the pushing field will contribute to an event, creating a sequential step that will affect the next event. Therefore, any movement produced by the waves will result in the experience of the flow of time by whatever is being affected by it.

If we reconsider the two mirrors, the field is moving the photon back and forth. An observer inside the spaceship looking at the photon will see the photon moving at the speed of light between both mirrors. However, the observer is also subject to the pushing field's velocity limitation. His movements, his observation, and his time are determined by the propagation of the field. If the two mirrors are moving in a certain direction at a high velocity, the observer is also moving at the same velocity. The photon takes longer to move between both plates, as it has to travel diagonally between the plates. The particles that make up the observer are also subject to the increase in the distance the particles have to travel. His movement will be delayed as a result of the velocity at which he is traveling. To the observer inside the spaceship, the photon still moves at the speed of light. Any time measurement device would also be subject to the same delay. Clocks would run slower. Everything in the spaceship is influenced by the pushing field and is subject to the increase in distance it has to travel because of the velocity at which it is traveling.

The observer outside the spaceship looking at the two mirrors would see the photon slow down as it bounces between both mirrors due to the extra distance the photon has to travel. The outside observer would also see the observer inside the spaceship slow down. For the second observer, the mirrors, the photon and the observer inside the spaceship would all be subject to time dilation due to the velocity at which they are traveling relative to him.

To expand on this idea, let us have the spaceship carrying the two mirrors circulate back to its point of origin. If we keep track of every time the photon bounces, there will be fewer bounces in the spaceship circulating around compared to the same experiment conducted in a stationary location. The two events originated from the same location, but because one is moving through space, its time flow is slowed down, as the particles have a longer distance to travel to get the same result. Individual clocks will run at different rates depending on the velocity at which they are moving. Time dilation is a result of the propagation of the pushing field through space at a set velocity. Moving closer to the velocity at which the pushing field propagates causes all events to take a longer time to occur. Time is derived from the sequence of events that can occur, and it will flow at the rate of the interaction with the pushing field. The pushing field theory duplicates the effect of Einstein's special relativity theory, but it also identifies the mechanism that produces the effect.

# WHAT IS THE ORIGIN OF TIME?

What is time? Everyone experiences the passage of time in one way or another. We normally understand time through some reference that measures it as it elapses. The concept of time has been contemplated by some of history's most important philosophers, and it is so fundamental to our universe that the study of time is considered crucial to the field of physics. Ideas about the passage of time have evolved over the years from Newtonian absolute time to Einstein's relativistic time. Time may flow differently depending on how fast you are moving or where you are located. Space and time have been linked in a fourth dimension called spacetime. Our understanding of time has grown from its first measurement with a sundial to the utilization of atomic clocks. Could there be a simpler explanation for the passage of time? Does time really exist, or is it just a construct of our imagination? Is there another possibility that exists that could explain the passage of time and duplicate the effects of gravitational time dilation?

The pushing field theory proposes that the passage of time is a product of a pushing field propagating through space at the speed of light and its resulting interaction with what occupies that space. Time does not flow on its own but is the result of the oscillation of the pushing field as it propagates. This idea means the propagating waves in the pushing field have two components: force and velocity. As the waves in the pushing field propagate through space, they are limited to the speed of light. This limitation creates the flow of time. What causes the limitation of the wave to propagate? What prevents the waves in the pushing field from traveling at an infinite velocity? Is something acting as a type of drag or resistance to the waves, creating the flow of time?

These questions can be a guide to an unbelievable answer. When the waves from the pushing field encounter matter, a force is exerted on matter from the waves. These waves are partially blocked from propagating any further. The effect of gravity is produced when the reduced waves emerge on the other side and the opposing waves, being stronger, push any other matter toward them. What happens to the component of time associated with the exhausted waves after the force component is released? Is it still present in the atom, and is it somehow dissipated back into space? Could this release of time add more of the "limiting factor" around the atom to cause extra drag on the other surrounding waves in the pushing field? Could this phenomenon be perceived as a slowdown in time in the area proportional to the distance it has originated from? Clocks would run slower near matter because of this "extra time" emanating from matter. This extra time does appears to radiate out of matter and can be compared to electromagnetic radiation. Time would emanate from matter, similar to how electromagnetic waves emanate from an antenna. This time radiation would slow down the progression of time when an area is exposed to it. Moving away from the source will increase the flow of time. This effect would not just be limited to matter, as it also would affect massless particles. Photons would also experience the extra time emanating from matter. In an attempt to devise an experiment to measure time radiation, it was realized that this phenomenon has already been observed with the index of refraction. It is established that light will slow down when traversing a medium, but the origin of this effect is still being debated. The index of refraction also causes light to bend when a photon traverses a different medium at an angle. It is understood that this effect is caused by a reduction in the photon's velocity when passing through a medium. Velocity is a measurement of distance over time. If the distance does not change, then time has to change. The velocity reduction of the photon is dependent on many factors such as the mass, the density of the material, and the chemical composition of the substance. There seems to be a correlation between all these factors and the reflective index. In other words, an increase in mass or density will result in more time radiation flux in that medium, which will decrease the velocity of the photon further.

The photon's velocity can be reduced not just inside the material but also at the exterior, near the surface. The effect will quickly decrease in intensity with increasing distance, as there is no medium to generate it. Surface time radiation can be observed when light passes through a slit. The time radiation from the wall of the slit will cause a photon near the surface to change direction and spread out.

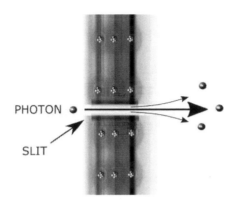

PHOTON

SLIT

Figure 26

*(Time variation near matter will cause the bending of light near the surface, resulting in light diffraction through a slit)*

This effect is not only limited to photons but extends to other particles, such as electrons. The behavior of the electrons can also be attributed to time variation. An electron travels faster outside a conductor then internally. The electron could be experiencing less time at the exterior of a conductor than inside the conductor. This is a different explanation from traditional physics, but this is where the questioning leads us. A different reasoning why electrons travel faster on the surface of a material than through it. The concept of time emittance can also be applied to the orbital lobs of an electron in an atom. When an electron is moving within an atom, the time radiating from the nucleus may not be uniform at this quantum scale. An electron cloud resembles the radiation pattern of an antenna. If the nucleus is somehow emitting time, similar to an electromagnetic wave, then at that close range it may resemble a similar lob pattern. The electron as it moves around the nucleus will be caught in this high intensity area of time and slow down considerably. This increases the probability of finding electrons in those locations. The probability associated with particles may have its origin in this time emittance. If an electron is stuck in a concentrated area of time, it would be more probable to find it there. The implications of time being emitted by nuclei are immense. There are many more phenomena to reconsider when contemplating the matter emittance of time, such as the polarization of light and Bell's theorem.

### Can this effect be from an opposing field?

To gain a deeper understanding of these phenomena related to time, let us go back to pushing field line oscillations and propagation. The pieces of this puzzle have started to culminate into a picture. It is now apparent that there is another field at play in respect to time and the pushing field. We can use the electric field to attempt to draw some conclusions. When an electric current flows in a conductor, a magnetic field is generated around the wire. A similar effect can be produced when a wave propagates through space in the pushing field lines. The oscillation of the pushing field lines is restricted through a restoring force. This force is produced by an unknown field that counteracts the oscillation of the pushing field lines, limiting its propagation to the speed of light. The oscillation of the field line propagates at the speed of light as a result of it being restricted from propagating any faster. Without the other field to limit the speed of the pushing field, its velocity would be infinite. Without this opposing field, time would not be a factor, and all events would happen instantaneously.

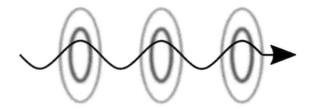

Figure 27

*(An opposing field excited by the oscillation in the pushing field opposes the propagation of the wave)*

The opposing field is produced perpendicular to the pushing field lines and opposes the propagation of the waves. The opposing field behaves similarly to an inductor working against the current flow. The pushing field and the opposing field are interconnected. There is a relationship that exists between the pushing field and the opposing field that can be formulated. In addition, when the energy of the wave is released as a force, the energy accumulated in the opposing field is released to additionally restricts the pushing field waves' propagation. The released energy by the opposing field will further restrict the pushing field wave velocity propagation in the surrounding area proportional to the distance where it is emitted. The additional restriction on the velocity of the propagation of the waves will be at its strongest at the point of origin and degrade over distance away from that point, which is the nucleus of an atom. The reduction in velocity will shorten the wavelength of the waves and result in a progressive compression of the pushing field waves as they approach matter. The reverse happens when the wave is moving away from matter. The geometry of the compression of the waves can be interpreted as the curving of space near massive objects, and the reduction of the velocity of the wave represents the slowing of the passage of time. This reproduces Einstein's space-time curvature. The interaction between the velocity of the pushing field and the opposing field brings about an interesting concept. Can the relationship between both fields be responsible for the arrow of time?

## A new understanding of time

Time may be the consequence of the pushing field propagating at the speed of light, and its effect is what we know of as the flow of time. Therefore, the flow of time originates from the restriction of the pushing field waves caused by an opposing field. In other words, the restriction of the propagation of the pushing field produces the effect of the arrow of time. The reduction in velocity of the pushing field has an impact on the flow of time, as the pushing waves cause events to occur. Increasing the potential of the opposing field causes additional restrictions on the pushing field's propagation. This, in turn, causes events to happen more slowly. This phenomenon is what we understand as gravitational time dilation. The reduction of the propagation of the pushing field waves decreases the wavelength and causes the pushing field waves to compress near matter. The phenomenon is not just restricted to large scales but can also be observed at the quantum scale. The reduction in the velocity of the pushing field wave can also be observed at the atomic scale, where it manifests as the index of refraction. The propagation of a photon originates from the pushing field waves. When the velocity of the pushing waves is reduced, the photon's propagation is also reduced. Near the nucleus of an atom, where the reduction of the pushing waves is at its greatest, the photons being pushed will be considerably slower. This phenomenon causes light to slow down or bend when approaching at an angle; however, in actuality, it is the pushing field lines that are being affected. Mathematician will call this curl in vector calculus, when a field rotates around a point. The oscillation in the pushing field lines will be bent or curl in a different direction depending on the interaction by this opposing field working against it.

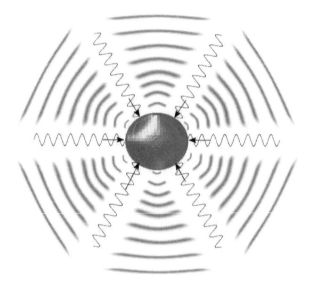

Figure 28

*(An opposing field emanating from matter reduces the velocity of the pushing field where it is being blocked)*

Not much is known about this opposing field that interacts with the pushing field to reduce its velocity and causes it to curl. It needs a name, so I have been calling this mysterious opposing field the "Kronos field", taken from Greek mythology, "the god of time". The closest comparison I can draw is that it resembles the magnetic field in some ways. We can deduce a few properties for this new Kronos field. A changing pushing field induces a

Kronos field. In turn, the Kronos field causes a reduction of the velocity of the pushing field and causes it to curl.

This relationship between both fields can be observed in two different phenomena. The first relationship is the universal velocity of light at which the pushing field propagates through space. As a result, the flow of time is produced. The second relationship is the interaction between matter and both fields that results in gravitational time dilation and the refractive index.

The speed of light is the balance between the pushing field oscillation wanting to propagate at infinite speed and the Kronos field reducing the velocity to the speed of light. This balance between both fields gives rise to the overall flow of time. It is the fundamental working of the arrow of time in the universe. The Kronos field opposes the pushing field wave propagation. If that did not happen, then all of history and future events would happen in the blink of an eye. The restriction of the propagation of the pushing field causes a slow progression of the waves' propagation through the universe. In turn, all action produced by the waves in the pushing field will take time.

Other observable phenomena are gravitational time dilation and the refractive index. When matter stops the pushing field waves, a force is generated that results in gravity. As a result, this increases the strength of the Kronos field. The Kronos field's strength is inversely proportional to the distance from the point where the force was exerted by the pushing field. Matter will block the pushing field oscillation, which results in a force. Therefore, matter produces an increase in the field strength of the Kronos field around it. The result will reduce the propagation of the pushing field waves near matter. We can describe the interaction of the Kronos field similarly to Maxwell's magnetic field interpretation. First, the Kronos field does not have any monopoles, so the first equations resemble Gauss's law for magnetism. The Kronos field will reduce the propagation of the wave in the pushing field. The Kronos field sets the velocity of the oscillation of the field lines. A representative equation should produce a value of the Kronos field that duplicates gravitational time dilation, as it will mimic the same effect. A Kronos field ($Kf$) equation will modified velocity of the pushing field ($c$) to become ($c'$). ($c'/c$) is then replaced by $n$, which is the index of refraction.

43

From this equation, an increase in the strength of the Kronos field will cause the pushing field to curl. This curling can be observed with the index of refraction (***n***) and gravitational time dilation (***c′/c***). The relationship between the Kronos field and the pushing field can be described as,

$$Kf = -\frac{F(Pf)}{r}$$

(6)

When a force from the pushing field ***F(Pf)*** is exerted, a Kronos field (***Kf***) will be produced at a distance ***r*** from the point where the force originated. The (***Kf***)will decrease at the same rate at which the circumference is increased. Thus, the curl of the pushing field (***Pf***) caused by the Kronos field can also be expressed as,

$$\nabla \times Pf = -\frac{\partial Kf}{\partial t}$$

(7)

An object such as a planet will cause the curl of the vector field to redirect toward the direction of the planet.

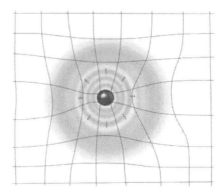

Figure 29
*(Representation of the curl in the vector field of the pushing field that resembles spacetime curvature)*

44

The pushing field and the Kronos field closely resemble the electric field and the magnetic field. Scientists have a good understanding of the electric and magnetic field, as evidenced by today's technological advancements. A similar outcome could be expected from an improved understanding of the pushing field and the Kronos field.

# KRONOS FIELD'S EFFECTS

## Light Diffraction

When a photon approaches matter at a 90-degree angle, it will experience time variation near matter equally along its width. When a photon approaches matter on an angle, it will be exposed to different strengths of the Kronos field across its width. The Kronos field causes curling in the pushing field that guides the photon. The curling of the pushing field will cause the photon to curve its trajectory to the stronger side.

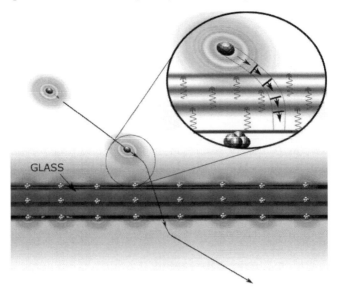

GLASS

Figure 30

*(When a photon approaches matter at an angle, the photon experiences the curl of the pushing field near the matter. This causes it to veer in its path toward the matter as it approaches and moves away.)*

If an observer could monitor the photon's behavior, he would see the photon approaching matter at a great speed, get caught in this "time molasse" and slow down. It would veer in its trajectory as it comes closer to the emission of the Kronos field. Time will take longer for the photon near matter compared to a photon in the vacuum of space. The Kronos field created by matter decreases with distance near the surrounding space.

The effect of the Kronos field will be reversed as the photon moves away from the object. The photon will veer to the side closest matter as it moves away from the object. This will curve the trajectory angle (refer to Figure 30). The final trajectory after it has passed through the glass will remain the same. The only evidence of the Kronos field would be the skewed path of the photon. This effect can be seen with a straw in a glass of water. We can see that the straw outside the glass does not line up with the straw in the glass of water.

Figure 31

*(Light passing through a glass of water demonstrates the skewed path of the photons)*

## Polarising Filters

Normal light is made of a number of electromagnetic waves propagating in different axes on the same path. With a polarising filter, we can remove or filter out electromagnetic waves of light of a particular axis. With two polarising filter positions at 90 degrees from each other, we can remove all the electromagnetic waves. Malus's law is a mathematical expression that describes how light passing through a polarizing filter can be varied by a secondary filter rotated at a certain angle. When a third filter is inserted at an angle of 45 degrees in-between the other two filters, something unexpected happens: there is an increase in light irradiance through all the filters.

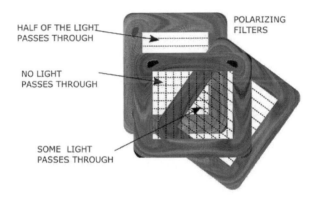

HALF OF THE LIGHT PASSES THROUGH

POLARIZING FILTERS

NO LIGHT PASSES THROUGH

SOME LIGHT PASSES THROUGH

Figure 32

*(The figures above illustrate three polarized filters. A filter positioned at 45 degrees in between the other two filters will allow some of the polarized light from the first filter to pass through the third filter)*

The polarizing lenses are made of slits that allow electromagnetic waves to pass from one axis and block the waves from the other axis. Each slit has matter on either side acting as a barrier to the electromagnetic wave that is not parallel to the slit. The barrier is made of matter and causes a Kronos field around it. The curl of the pushing field caused by the Kronos field emanated by the filters slits twists the electromagnetic wave. This allows some of the waves to pass through the third filter. The electromagnetic wave that is exactly parallel to the slit will pass through with no problem. The electromagnetic wave that is at a slight angle to the slit will follow the time curvature caused by the Kronos field near the barrier. This will "twist" the electromagnetic wave and allow it to pass through the next set of slits.

The Kronos field creates a curvature of time layered around the slits which photons are exposed to that shifts their polarization.

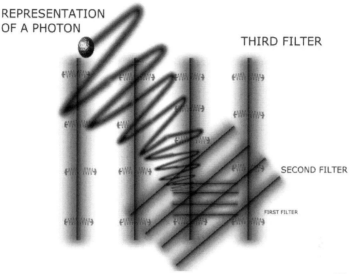

REPRESENTATION OF A PHOTON

THIRD FILTER

SECOND FILTER

FIRST FILTER

Figure 33

*(Electromagnetic waves are curled as they pass through the filters and changes the direction of the polarity)*

The same effect could explain light dispersion through a slit. Light as it passes through a slit will follow the curvature of time near the edges of the slit and get re-directed on a new trajectory. Light passing close to the surface of an object guided by the pushing field will curl when exposed to a Kronos field and follow a new path.

Using a Kronos field to curl the pushing field near matter is an interesting concept that provides an explanation for the different behaviours of light. This effect may not be limited to photons and may be seen with other particles. The effect of time variation experienced by the photon should be experienced by other particles such as the electron. The electron is a negatively charged particle, which causes the electron to be attracted to the positive nucleus of an atom. The electron is anchored to the nucleus and is in a constant state of movement. This will play out differently than the photons in respect to the curling of the Kronos field, which may increase the probability of detecting it.

# QUANTUM PROBABILITY

When a photon interacts with an atom, it is only for a fraction of a second as it passes through the atom. An electron tethered to the nucleus of an atom by its negative electric charge will spend a considerably longer amount of time interacting with the atom. The electron will occupy certain regions around its nucleus. These regions around the atom are the most probable places to find the electron. Could the probability of the electron's position be determined by the Kronos field? The electron's movement occurs as a result of the pushing field moving it around. When the electron moves near the atom, it will pass through areas with variation in the Kronos field. The pushing field's velocity is reduced with an increase in the Kronos field. This means the electron's velocity will also be inversely proportional to the increase in the Kronos field. If the electron spends more time in certain areas as a consequence of the Kronos field, then it is more probable that the electron will be found at those locations. This means the Kronos field is partially responsible for the atom's structure. There are a number of other factors at play, but what I am suggesting is that a portion of the probability of the electron's position around the nucleus may be determined by the variations in the Kronos field created around the nucleus.

If there was more "time" due to an increase in the Kronos field, what would it look like? Is it possible that the electron orbitals provide a clue to this phenomenon? The area the electron occupies around the nucleus resembles a partial magnetic field, similar to a magnetic field around a magnet. The subatomic particles within the nucleus may be emitting Kronos fields, and the positions they take within the nucleus create these geometric areas around the nucleus with higher intensities.

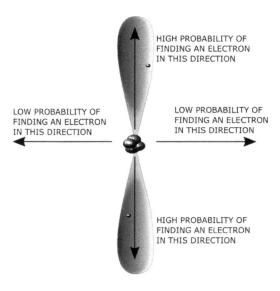

HIGH PROBABILITY OF
FINDING AN ELECTRON
IN THIS DIRECTION

LOW PROBABILITY OF
FINDING AN ELECTRON
IN THIS DIRECTION

LOW PROBABILITY OF
FINDING AN ELECTRON
IN THIS DIRECTION

HIGH PROBABILITY OF
FINDING AN ELECTRON
IN THIS DIRECTION

Figure 34

*(Example of Electron Cloud Probability)*

The Kronos field may be generated by the nuclei, similar to small magnet. The addition of nuclei will change how these "magnets" are pointing, which could change the structure of the atom. The electrons moving around the nucleus will be caught in a concentrated region of field lines from the Kronos field. Once an electron occupies a region, its negative charge repels the other electron from that area. Other electrons will position themselves in other areas around the nucleus with the least negative repulsion and highest concentration of Kronos field lines. All the combination of the electrons' positions will form the atomic structure.

THE CONCENTRATION OF KRONOS FIELD LINES CAUSES
ELECTRONS TO SPEND MORE TIME IN THIS AREA
RESULTING WITH A PROBABILITY CLOUD

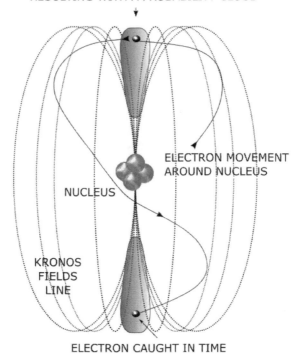

Figure 35

*(A higher intensity of the Kronos field slows down the electron's velocity, resulting in the probability of the electron cloud)*

The Kronos field will decrease in strength at the outer layers of the atom and combine with other fields from nearby atoms to create a uniform region extending outwards from those atoms. An electron at the outer layers of the atoms will move around easily but will still gather in the concentrated areas of field lines from the Kronos field.

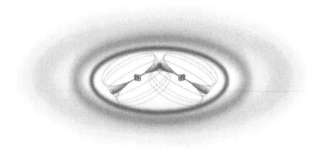

Figure 36

*(The illustration represents the distribution of the Kronos field produced by a couple of atoms)*

This is an interesting idea, but is it really the case? I would not be surprised to find that the Kronos field does in fact play a role in the atomic structure. The Kronos field and the pushing field are responsible for some of the behaviors of subatomic particles; there is currently support for this in quantum mechanics, or least a variation of it, which is discussed in the next chapter.

# PILOT WAVE THEORY

The pushing field theory may offer an alternative hypothesis for gravity and inertia, but can it do the same for other areas of physics? For example, the strong nuclear force, which is a quantum force, could be just one of the effects of the pushing field. What else should be considered in the quantum realm? Is there an existing model that would compliment the pushing field theory? Quantum mechanics has been the leading contender for explaining this microscopic world and has proven to be the most accurate model for predicting the behavior of particles. One variation of quantum mechanics called Bohmian mechanics has the same fundamental equations but maintains the classical view of physics. These theories have originated from a mathematical extrapolation of the Schrödinger equation in an attempt to explain observable phenomena from the double-slit experiment. Although the uncertainty model evolved into quantum mechanics, there has been progress in the classical approach that is reviving Bohmian mechanics.

Bohmian mechanics or (de Broglie–Bohm theory), also known as pilot wave theory, has been around since 1927, when Louis de Broglie first proposed that a particle is accompanied by a pilot wave. He derived this theory from the Schrödinger equation, the same equation that quantum mechanics theory was derived from. He postulated that the Schrödinger equation can be broken down into two separate equations. One equation is for the wave function of the particle, and the other is the wave that guides the particle. Does this not sound exactly the same as a photon with carrying frequencies in the pushing field? In this model, the particle has a defined path. If you could determine the starting condition of a particle, you could predict its outcome. This classical approach received considerable resistance when it was first proposed. The revolutionary physicist who favored quantum mechanics did not agree with this approach. Broglie abandoned this idea after receiving the negative feedback and continued to work on quantum mechanics. David Bohm, twenty-five years later, continued his work to complete the theory. This classical view on the universe predicted how particles behave regardless of the observer. The supporters of pilot wave theory argued that it is the ignorance of the starting condition that gives it uncertainty. It is the unknown starting conditions, referred to as hidden variables, that cannot be measured without changing their value, which will affect the outcome of the particle. This is the argument used to explain the double-slit experiment. The observer does not know all the starting conditions of the particle, and, therefore, the particle appears to behave randomly. But, in fact, the particle has followed a well-defined path, eliminating the uncertainty. This path followed by the particle comes from a guiding wave that is separate from the particle. This wave exists, even if there is no particle present. The interaction between the waves and particles produces the interference pattern observed in the double-slit experiment.

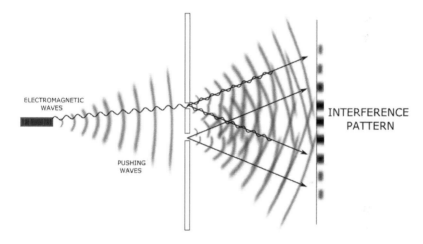

Figure 37

*(Demonstration of the double split experiment producing an interference pattern)*

Pilot-wave theory duplicates all quantum mechanics phenomena, for example, quantum tunneling and particles entanglement. Despite this, physicists have moved away from the theory, and some have even taken a stance against it. The existence of a wave function separate from the particle has left physicists perplexed. What is this mysterious wave guiding particles on their path? Are there empty waves roaming through space? Physicists are not seeing the potential of de Broglie's pilot wave theory. These waves propagating through space are not just a mathematical explanation for the behaviours of particles. These waves could be responsible for fundamental forces in the universe. As described in the previous chapters, these pilot waves could be the ripples propagating through the pushing field. These waves would produce gravity, inertia, and propagate photons through space. These invisible waves must be present in a field moving across the universe like ripples moving on water. This field is capable of moving particles, and it also guides them to their destination.

## Bosons and Fermions

Here is an interesting idea that furthers this argument. Can the pushing field be fundamentally responsible for the characterisation of particles? Particles behave differently in the pushing field depending on their wave function and how the wave function interacts with the field. One earlier example described the photon's ability to move at the velocity of the pushing field. This ability may come from its wave function being identical to the pushing field. What happens when the wave function does not "match" the pushing field oscillation? These dissimilar waves may interfere with each other and create resistance between them. The resistance of a wave function to the pushing field's oscillation ends up as a force working against it. Particles are classified into two groups: fermions and bosons. One distinguishing aspect of the two groups is the symmetry of their wave function. A boson wave function is symmetric, while fermions are antisymmetric. Fermions oppose the pushing field oscillations in all directions, and the end result is a force exerted on the particles. The resistance to the pushing field waves could also give the property of mass to the particle. In effect, the constant force exerted from the pushing field will vary depending on the particle's resistive value. This resistive value is, in a way, the atomic mass of the particle. The greater the resistive value, the greater the mass.

The boson particles, which the photon belongs to, do not resist the oscillation of the pushing field from all sides. This would allow the pushing wave to pass through them without any resistance from at least one direction. These particles will be pushed by the opposing waves at the velocity of the pushing field, which is the speed of light. The path particles take depends on the forces exerted on them as they pass through the field.

The probability of the trajectory of the particles depends on the starting condition, the forces at play, and the influence of other fields on the path. This new understanding would explain the results from the double-slit experiment. The waves that guide the photons on its trajectory will pass through both slits, even though the photon only passes through one. The pushing waves will curl when exposed to a variable Kronos field at the slit and will be redirected to diverge on the other side of the slit. The waves that started with the same phase and frequency will cause an interfere pattern on the other side of the slits. A particle being guided by the wave will be subjected to the wave's own interference. Every particle has a specific wave function and therefore interacts differently with each wave frequency of the pushing field. Although there are a number of wave frequencies being affected by the slit, only one frequency will "carry" and produce the interference pattern. Every particle will have its own interference pattern, which depends on the frequency that is carrying the particle in the pushing field. The result of the double-slit experiment also suggests that the pushing field has a number of different frequencies propagating as waves. Each one interacts differently with the wave function of the particles. The frequency of the pushing field is identical to the wave function of the particles, and both waves are entangled at the moment of creation. The particle will follow the pushing field wave until an event breaks the bond between them. The waves from the pushing field are susceptible to miniscule changes, making them difficult to measure. As soon as the wave is exposed to a measurement device or is redirected for a measurement, it is no longer part of the same wave. Such an experiment would produce data that could be misinterpreted.

A great deal of research can be done in regard to quantum mechanics and the pushing field theory. For example, how does the Kronos field affect subatomic particles? Why is the field not uniform? There is so much more to investigate that this topic deserves a book of its own dedicated to it. I wanted to demonstrate that a variation of quantum mechanics does agree with the pushing field theory. I believe the pilot wave theory provides additional evidence for the existence of the pushing field.

# CENTRIPETAL FORCE

The centripetal force or centrifugal force has long been thought of as a fundamental phenomenon; however, perceptions on the origin of this force have changed over the years. In the early 1600s, Christiaan Huygens was the first to use the words "centrifugal force" to describe this force as acting outward. Isaac Newton used the words "centripetal force" to describe this phenomenon as a force acting inward. Both views differ in their perception of what causes this effect. It was thought that the centrifugal force is a force exerted outwards on an object as it changes direction. This view is now believed to be incorrect, and the centripetal force is used to describe how an object behaves as it accelerates toward the center of a circle. By the 1900s, the debate on this phenomenon quieted with the "reference frame" in which it is considered, and there have been no considerable advancements on this phenomenon since then.

However, in an attempt to account for the existence of the pushing field, it was discovered that this force could be the result of two individual forces. The centripetal force could be caused by an object's inertial acceleration on its new trajectory and the inertial deceleration of the same object from its old trajectory. These two combined effects on the object create the phenomenon we know of as centripetal force.

The inertia of an object can be seen when it accelerates or when the object changes direction without reducing its velocity. If that object does not take a circular path, it will have to decelerate to a complete stop in its current vector, turn to its new trajectory and accelerate on this new vector. As a result, the object will experience inertia as it stops and as it accelerates. What would happen if the object started to accelerate in a new direction as it decelerates from its previous direction of travel? Do these two inertial effects just disappear, or do they combine to become a single phenomenon we know of as the centripetal force? How can we consider the contribution of each individual phenomenon as a single effect and compare it to the centripetal force? Let us consider inertia as a product of the pushing field as theorized by the pushing field theory. The differential force produced by the pushing field as it acts on an object should be perpendicular to each other. The two forces are assigned to different axis as the object moves in a circular motion. Both force combine should reproduce the same force we know of as the centripetal force.

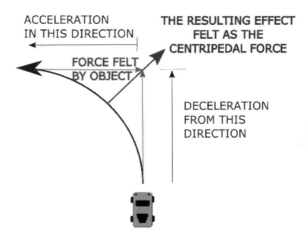

Figure 38

*(Simplified illustration demonstrating inertia as a force reproducing the effect of centripetal force as an object moves in a circular motion)*

## CALCULATION

The force required to accelerate an object is simple and can be calculated using F=ma. F=the force applied on the object, a=the acceleration of the object, and m=the inertial mass of the object. If we know the inertial mass and the acceleration, we can calculate the force required to move that object and reach that acceleration. Let us consider a vehicle negotiating a 90-degree radius corner. If a vehicle was to change its direction of travel without changing its velocity, then it must do so in a circular motion. The force of friction generated by the wheels pointing in a different direction causes this change in direction. As a result, acceleration will be generated from that change in motion. The formula **F=ma** can be modified to understand the inertia on the vehicle as it progresses on the curved path. This change in direction on a circular path can be considered as a transition of two linear velocity vectors at 90 degrees from each other (refer to Figure 38).

The first linear velocity vector is the direction in which the vehicle was travelling before taking the curved path, and the second linear velocity vector is the direction the vehicle will be taking after the curved path. The resulting force on the vehicle from the deceleration and acceleration of the two linear velocity vectors should reproduce the same force on the vehicle that we can calculate for the centripetal force.

To consider the deceleration of the vehicle as a linear velocity vector from one direction and the acceleration of the vehicle as another linear velocity vector from another direction, the forces acting on the vehicle must be considered to be a variable throughout the curved path. As the object deviates from its original linear velocity vector and embarks on its new linear velocity vector, it does so in a circular motion. The progression of the object on the curve as it deviates from the original linear velocity vector can be expressed as an angle in radians.

The forces resulting from the change of each linear velocity vector will be independent from each other and will vary through the curve. The rate at which the vehicle decelerates and accelerates will depend on which point the vehicle is located on the curved path. When the vehicle first starts on the curved path, it will experience a small deceleration from the change of the linear velocity vector it was previously on. As the vehicle continues on its course, the deceleration will gradually increase to the maximum value as it reaches the end of the curve. The value of the force felt by the object caused by the deceleration can be calculated using F=ma. To consider the amount of deceleration occurring as a result of the curve, the radian angle is used in the calculation. The expression is modified to include the angle progression in the curve and is shown below.

$$Fd = m * d(axis) * (rad) \tag{8}$$

$F_d$ is the linear velocity deceleration force felt by the object, **m** is the inertial mass of the object, $d_{(axis)}$ is the deceleration of the object on the vector on which it is considered, and **(rad)** is the radian angle at which the object has progressed on the curved path. The first step is to find the rate of the decreasing velocity as the object decelerates on the original vector on which it was traveling. When the object is decreasing its linear velocity from this vector, it is doing so on a curved path. The deceleration from this vector $d_{(axis)}$ can be calculated as follows:

$$d(axis) = (v\ final - v\ initial)\ /\ t \qquad (9)$$

The vehicle's velocity before the 90-degree turn will be ($v$ initial), and ($v$ final) will be the linear velocity from this vector at the end of the 90-degree turn, which will be zero. There will no longer be any momentum from the vehicle in that direction. It should be noted that a 90-degree turn was utilized to maximize the effect and simplify the calculation. To find the time (**t**) the vehicle has taken to complete the 90-degree turn, we simply use the circumference of a circle ($2\pi r$), divide it by four to get the distance of the curve and divide the result by the velocity ($v$).

$$t = (\ 2\ /4\pi r\ /v) \qquad (10)$$

The result of ($^2/4\pi r$ /**v**) can then be utilized for the calculation of the $d_{(axis)}$. Once the distance $d_{(axis)}$ is calculated, it can be used in the formula $F_d = m*d_{(axis)}*(rad)$ to get the force felt from the deceleration of this particular linear velocity vector. The radian angle can provide incremental angles on the curved path to analyze the forces. Each incremental radian angle is equal to an amount of distance travelled on the curved path. The force from the linear velocity deceleration can be plotted to visualize the entire effect. For the calculation, a vehicle weighing 1721 kg is utilized for the mass, with a radius of 250 meters for the curved path, and a velocity of 27.78 m/s (100 km/hr) for the velocity. The calculation for $d_{(axis)}$ is as follows:

Time traveled for the vehicle on the curved path

$$(^2/4\pi r\ /\mathbf{v}) = (2\pi*250\ m\ /\ 4)\ /\ 27.78\ m/s = \mathbf{14\ seconds}$$

Deceleration of the object on the curved path ($d_{(axis)}$)

$$d_{(axis)} = (27.78 \text{ m/s} - 0 \text{ m/s}) / 14 \text{ seconds} = \mathbf{1.9842 \text{ m/s}^2}$$

Using the formula $\mathbf{F_d = m * d_{(axis)} * (rad)}$, it is possible to calculate the force produced by the deceleration of the vehicle from the original vector. Multiplying the mass (1721 kg) and the axial deceleration (1.9842 m/s²) by increments of 3.75 degrees (0.065625 radians) produced the results shown in Figure 39. Each increment angle translates to a distance traveled by the vehicle on the curved path of 16.35 meters. The force from the deceleration acting on the object was plotted on a graph to demonstrate the forces felt by the vehicle and its occupant as it progresses on the curved path. The graph shows the behaviors of this force as a gradual increase of the deceleration force on that object.

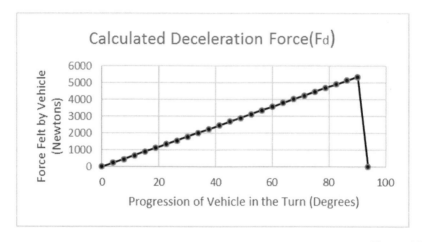

Figure 39

*(Graphical result of the force produced by the deceleration of the vehicle from the linear velocity vector as it progresses in a new direction)*

From the results shown in Figure 39, the force acting on the vehicle starts with a low value at the beginning of the curve and gradually increases to its maximum value by the end of the curve. This force will be acting on the vehicle and its occupants in the same direction as its original path of travel. Once the vehicle has reached the end of the curved path, the effect stops, as there is no longer any momentum left from the initial trajectory.

The same process is used to obtain the force encountered when accelerating in the new linear velocity vector, but there is one main difference: The acceleration ends when the object is no longer on the curved path. Therefore, the calculation has to consider the radian angle as a decreasing value toward the end. In a 90-degree change in direction, the object begins with the maximum acceleration as it begins on the new vector and gradually decreases in strength as it approaches its final velocity on that trajectory. Once the object has completed the circular motion, the effect stops, and there is no longer a force acting on the object. The calculation is similar to the deceleration, but the deviation angle is from the new vector. The end of the curve is the point where the deviation begins with zero degrees, and the beginning of the curve is the maximum deviation with 90 degrees. The same formula is used with the following changes.

$$Fa = m * a(axis) * (rad) \qquad (11)$$

$Fa$ is the acceleration force of the object on that new vector, $m$ = mass of the object, $a(axis)$ = acceleration of the object on this vector, and $(rad)$ = radian of the object angle on the curved path. To obtain the acceleration on the new vector, the following formula is used.

$$a(axis) = (v\ final - v\ initial) \ / \ (r^2/4\pi r\ /v) \qquad (12)$$

The values obtained from the calculation are plotted in Figure 40.

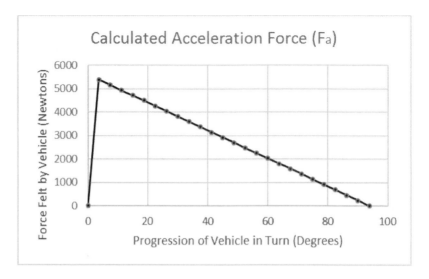

Figure 40

*(Graphical result of the force produced by the acceleration of the vehicle in the new linear velocity vector as it progresses in the new direction)*

The results shown in Figure 40 demonstrate the behavior of the force acting on the vehicle as it accelerates on its new vector. The force is at its strongest when it first starts the curve and gradually decreases as it accelerates on its new vector. The force stops when the vehicle has completed the turn and all its momentum is heading in the new direction.

Both forces, the force experienced by the vehicle's deceleration from its past vector and the force felt by the vehicle's acceleration in the new vector, combined will result in the centripetal force felt by the vehicle. The combination of the two varying forces working 90 degrees from each other will result in a constant force felt in a direction away from the center of the curve, making it appear as an acceleration toward the center.

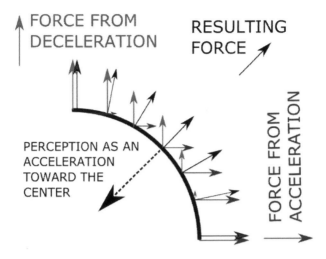

Figure 41

*(Illustration of the deceleration and acceleration forces acting within a curved path to reproduce the centripetal force)*

The sum of the net force from the deceleration and acceleration will be the same value throughout the radius. As the force from the deceleration increases in the radius, the force for the acceleration decreases, but the resulting net force will remain the same with a value of **5361.48 Newtons**. This is plotted on the graph shown in Figure 42.

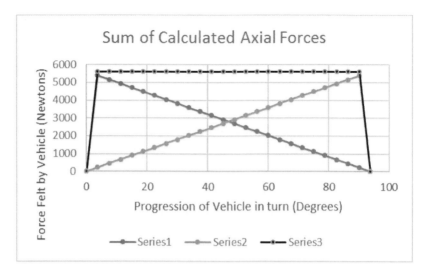

Series 1 – Calculated force from Acceleration (Fa), Series 2 - Calculated Force from Deceleration (Fd), and Series 3 – Sum (Fd + Fa) Resulting as Centripetal Force.

Figure 44

*(Graphical result of the calculation of the force from deceleration and acceleration to reproduce a constant force rotating toward the center of a circular path)*

The net force resulting from the acceleration and deceleration working in combination on the object duplicates the behavior of the centripetal force with a similar strength. The net inertia will act in the direction perpendicular to the radius throughout the curve. The direction in which the net force is acting on the object will make it appear as if the object is accelerating toward the center point, producing the effect we refer to as the centripetal force. In comparison, the calculated value of the centripetal force using the formula ($Fc=mv^2/r$) produces a value of

**Centripetal Force (Fc)** = 1721 kg * (27.78 m/s)$^2$ / 250 m = **5312.57 Newtons**

To further explore this idea and confirm the calculation, an experiment was devised to test this hypothesis.

## MEASUREMENT APPARATUS

The device consisted of two accelerometers positioned perpendicular to each other to measure the acceleration and deceleration components on their respective vectors. The apparatus had to maintain the accelerometers on their original vector as they maneuvered through the curved path. A third accelerometer was used to measure the centripetal force as a control. The apparatus was mounted on a vehicle that travelled a section of a highway with a 90-degree curve. Two Samsung cellphones were used as accelerometers.

 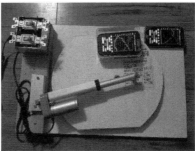

Figure 43                                    Figure 44

*(Pictures of the apparatus with one cellphone mounted on the turntable and the other mounted to its base. Figure 43 shows the apparatus in the starting position, and Figure 44 shows it after the device has been activated.)*

An actuator was connected to a turntable, which rotated the accelerometers to maintain their axial position within the turn (see Figure 43 and 44). The actuator was activated at the beginning of the curve to rotate the turntable at a fixed speed, matching the vehicle speed on the curved path. A variable power supply was utilized to adjust the voltage to the actuator as a speed control for the turntable. The power supply voltage for the actuator was adjusted so that the turntable took the same amount of time to rotate 90 degrees as it took the vehicle to travel the 90-degree turn. The "Accelerometer Meter" app from Keuwlsoft was used to record the values obtained from the experiment. The recorded values were then transferred to an Excel spreadsheet. One cellphone measured both the acceleration and deceleration on the curved path, and the other cellphone measured the centripetal force. To perform the experiment, a section of highway with two 90-degree radii were used as shown in Figure 45.

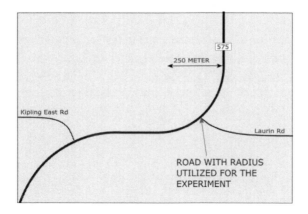

Figure 45

*(Illustration of the curved path on which the experiment was performed with a 250-meter radius)*

The operator of the vehicle maintained a constant speed throughout the experiment. A second operator activated the switch of the power supply at a pre-determined location at the start of the radius. The actuator, once activated, completed a full 90-degree rotation of the turntable, matching the time it took to complete the curve on the highway. The position of each accelerometer relative to the curve is shown in Figure 46.

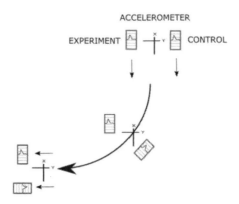

Figure 46

*(Demonstration of the position of the cellphone as the vehicle started on the curved path)*

**RESULTS**

The experiment reveals similar values and is comparable to the calculations. The deceleration from its initial vector started with a lesser value and increased to its maximum value at the end of the curve. The acceleration on the new vector also occurred as expected. It started with a high value and decreased to a lesser value by the end of the curve. Both values added together reveal a similar value as that obtained by the control accelerometer, which measured the centrifugal force.

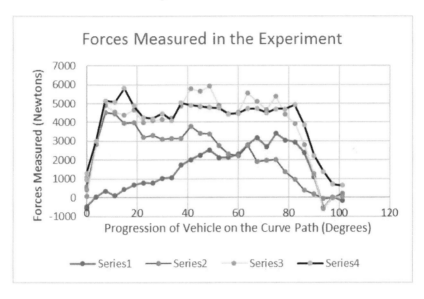

Series1 – Force from Deceleration Measured (Fd), Series2 – Force from Acceleration Measured (Fa), Series 3 – Sum of Forces from Deceleration and Acceleration Measured (Fd + Fa), and Series4 – Centripetal Force Measured (Fc)

Figure 47

*(The measurement results from the experiment)*

The experiment provides comparable results as those from the calculation. Figure 47 demonstrates the forces acting on the vehicle as it negotiated the turn. The vehicle's inertia was measured as two separate components acting together on the object; the measured axial value also collaborated with the control measurement of the centripetal force. Further testing should be conducted with more precise equipment and with more control conditions. For example, to improve the experiment, more precise accelerometers could be mounted on a rotating device traveling on a fixed track. This would provide a level accelerometer with a true radius. Overall, however, I believe the experiment was a success. The experiment could also be expanded to examine other phenomena. First, we could explore the forces at play when a radius is changed as an object is taking the curved path. Second, we could explore the forces acting on an object as it accelerates or decelerates while taking a curved path. The outcomes of these experiments could be calculated and predicted using this method. This could shed new light on rotational inertia and angular momentum.

# GYROSCOPIC EFFECT

Any change in motion will result in a force from the pushing field that counters that motion. A change in motion on the same plane from an acceleration will result in a force in the opposite direction. The change in motion on two separate planes (an acceleration in the Y-axis and an acceleration or deceleration in the X-axis) will result in two forces on each axis that combine to act as one force known as the centripetal force. When in constant circular motion, this force becomes a constant force as the object rotates. For example, a spinning wheel will have a constant acceleration force, as the particles that constitute the wheel are in a constant state of acceleration in a new direction, while at the same time they are in a constant state of deceleration from the previous direction. This is the same fundamental principle of a doppler shift resulting from the change in motion with respect to the wave from the pushing field. Multiple changes at the same time will result in multiple forces acting at the same time. Our universe is represented by three special dimensions, and different phenomena are associated with each plane. The first phenomenon is a change on a single plane and is known as acceleration or deceleration. The second phenomenon is a change on two planes simultaneously resulting in the centripetal force. The third phenomenon is a change on three planes and is what I call the gyroscopic effect.

When an object rotates on itself, the rotation has an amplification effect. The force produced from the acceleration on the third plane will be multiplied by the object's rotation. A wheel with no rotation will not produce this effect. When the wheel has a rotation, every time the particles that constitute the wheel pass by the area where the change is occurring, it will experience an acceleration and, therefore, a force resulting from that acceleration. This force will be felt by the particle not just once, but every time the particles passes over the same area where the force is being applied.

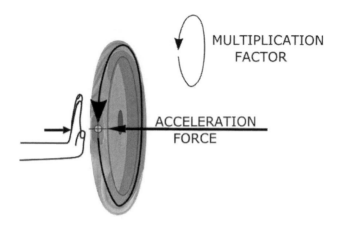

MULTIPLICATION
FACTOR

ACCELERATION
FORCE

Figure 48

*(The illustration demonstrates how when a force is applied to cause a movement of the wheel, a force from that acceleration will result that counters the movement. This force is multiplied by the rotation of the wheel.)*

The number of times the force is felt in a set period of time will be the total force felt at that location as the wheel is accelerated perpendicular to the rotation. If the force is applied off the center of gravity of the wheel, then the wheel "should rotate" around the center of gravity. This should cause an acceleration not just where the force is applied, but on the opposite side of the wheel, as the wheel "should rotate" around the center of gravity. This movement of the wheel will also cause an acceleration on the other side of the wheel. This will produce another force that will also be multiplied, but in the opposite direction. Both multiplied forces working together will result in the wheel moving in the direction of the applied force. This is the behavior of a rotating object in space with no restrictions. A detailed analysis is needed to truly understand the gyroscopic effect. Many factors play a role in the behavior of the wheel, and all conditions must be analysed. The first condition to consider is what other force will come into play when a change in the wheel's motion occurs. The wheel will behave differently depending on whether the wheel has restrictions on its movement or not. A rotating wheel in microgravity will do different things than a rotating wheel attached to a bicycle.

A free-floating rotating wheel in space will keep its orientation, and the force applied to the wheel will cause the entire wheel to move in the direction in which the force was applied. The force produced by the wheel as it rotate on its center of gravity will move the wheel in the same direction as the force being applied. When the wheel has restrictions on its movement, those restrictions will cause a different effect. Handles are added to the wheel to assist with the explanation of the effect.

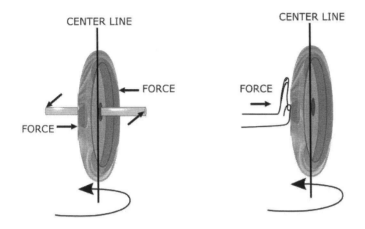

Figure 49

*(Demonstration of how a force applied to a non-spinning wheel away from its center of gravity would cause an axis of rotation.)*

The wheel is given a spinning rotation. The handles are used to push and pull on either side of the wheel. The push/pull action on the wheel will cause a rotation on its axis in the direction of the push. This can be considered the same as pushing on the side of the rotating wheel to cause a rotation around the center of gravity. The wheel will be changing its movement perpendicular to its rotation. The push/pull on the wheel causes an axis of rotation on the wheel. The wheel should be rotating around the center axis. The push on the side of the wheel will accelerate the side of the wheel. This acceleration will generate a force. The force that acts on the wheel is not even. The entire wheel is not accelerated evenly in respect to the pushing field acting on the wheel. At the location where the wheel turns on its axis (top and bottom), there is no acceleration. At that location, there is no movement on the third axis, and, therefore, no force is produced at that location.

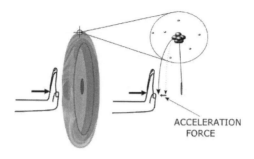

Figure 50

*(In a wheel with a rotation, the atoms that constitute the wheel will be accelerated perpendicular to the wheel as the atom moves away from the axis of rotation. This action will cause a force that will be multiplied by the wheel's rotation.)*

Particles (or molecules) that constitute the wheel progress from the top of the wheel to the furthest point away from the axis, which is halfway down the wheel. The particles (molecules) are gradually accelerated because of the rotation caused by the push on the wheel. As a result, a force is generated on the wheel. At the furthest point away from the axis of rotation, the force will stop. At that location, the force caused by the acceleration changes direction. From that point on, the particles are accelerated in the opposite direction. The particles are now moving back to the axis of rotation.

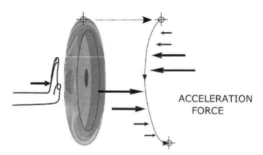

Figure 51

*(Once the atoms have reached the furthest point away from the axis of rotation, the force is reversed in the opposite direction, as the atoms are making their way back to the axis of rotation.)*

The change in direction causes the acceleration to be in the opposite direction. Therefore, the force created will also be in the opposite direction. Both forces (top and bottom) produced from the push on the wheel will result in a torque of the wheel. This torque will cause the wheel to move perpendicular to the initial push pull on the wheel.

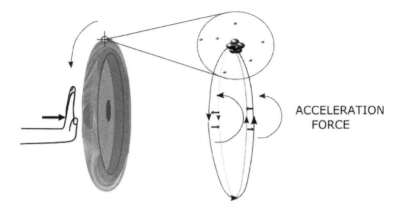

Figure 52

*(The forces are in an opposite direction and produce a torque on the wheel. The far side of the wheel also produces the same effect but in the opposite direction. The forces produced by either side of the wheel are pointing in the same direction.)*

These forces acting in different directions, at different locations, are amplified by the wheel's rotational speed and produce the unexpected behavior of the wheel. By pushing and pulling on the handles, we expect the wheel to twist. But, instead, the wheel will twist perpendicular to the force that was applied to the wheel.

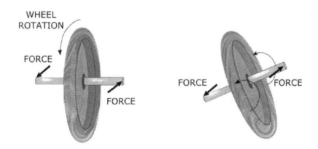

Figure 53

*(The wheel will twist perpendicular to the direction of the force applied to the wheel, producing an unexpected behavior.)*

## Spinning Tops

A spinning top is not much different than a spinning wheel on its side. The wheel is supported by one of the handles, while the other handle is eliminated. Gravity is doing the push/pull on the wheel. As soon as there is acceleration at any particular point on the spinning top, there will be a force caused by that acceleration. This force is multiplied by the rotation of the spinning top. The acceleration force multiplied by the rotation of the spinning top will push back on the spinning top. This force is continuously keeping the wheel from falling over. The force produced by the acceleration keeps the spinning top from falling over.

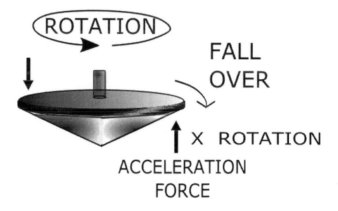

Figure 54
*(A spinning top will maintain its balanced position as long as it has rotation.)*

Gravity is a factor that has to be accounted for when considering a rotational object. If gravity is the force acting on the object, then another phenomenon called precession will take place.

## Precession

A wheel's rotation will cause an amplification of the acceleration force acting on it. By pulling on the handle, the wheel produces a torque perpendicular to the handles. When only one handle is supporting the wheel, then the acceleration forces that cause the torque will cause the wheel to rotate around the pivot point. A rotating wheel positioned on a pivot point accelerates downward due to gravity. As the wheel falls to one side, it causes an axis of rotation on the wheel. This is the same scenario as previously explained when pushing and pulling on the handles of a rotating wheel. Instead of the wheel rotating horizontally, it is rotating vertically.

76

The torque produced is perpendicular to the direction of gravity that is pulling down. The torque will cause a rotation of the wheel around the pivot point. At the same time, the force of acceleration acting on the wheel keeps the wheel from falling over.

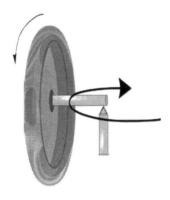

Figure 55

*(The wheel will rotate around a pivot point as a result of the torque created by gravity.)*

Precession has the same principles as the gyroscopic effect, but gravity does the pulling. The force amplified by the rotation of the wheel will hold the wheel up. It is the acceleration of the wheel as it falls over that prevents the wheel from falling over.

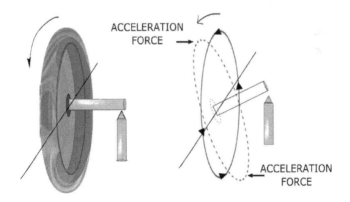

Figure 56

*(The wheel is held up by the force of acceleration caused by gravity trying to pull the wheel down.)*

# A NEW EXPERIMENT TO CONSIDER

Early in 2008, Louis Rancourt published a paper on an experiment he conducted that investigated how electromagnetic waves interferes with gravity. At that time, he concluded that gravity seemed to behave as a repulsive force. His experiment was simple and involved a pendulum on a string with a wide-angle laser beam aimed to one side of the pendulum. He expected the pendulum to move away from the laser or not to move at all. Specifically, he expected the laser to interrupt the gravity field from nearby objects, causing the pendulum to move away from the laser. To his amazement, the pendulum moved closer to the laser. He repeated the experiment a number of times and always got the same result. To remove the possibility of interference from air currents, he put the pendulum in a box; however, the same result was observed. He decided to publish his results in the journal Physics Essays.

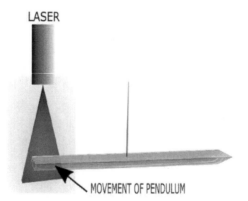

LASER

MOVEMENT OF PENDULUM

Figure 57

*(Figure 57 illustrates Louis Rancourt's experiment using a laminar laser beam beside a suspended mass. The suspended mass moved toward the laser beam when turned on.)*

*Simple rotational pendulum to verify how light blocks gravity, Applied Physics Research; Vol. 7, No. 4: 2015, ISSN 1916-9639, E-ISSN 1916-9647

Rancourt did not stop his experimenting there, as he continued his research by placing a 35,000-lumen light source above a weight positioned on a high-precision scale. He observed that the weight lessened over a period of four hours. The weight gradually returned to normal after the light was turned off. The results are shown in Figure 58.

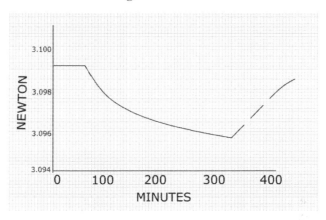

Figure 58

*(The results of his experiment demonstrate a lessening of the weight of the object over a period of time.)*

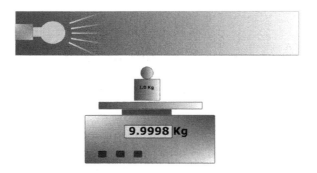

Figure 59

*(The illustration above demonstrates a representation of the experiment, but it is not an accurate depiction)*

He published his findings: **\*Further Experiments Demonstrating the Effect of Light on Gravitation, Published by the Canadian Center of Science and Education, Applied Physics Research; Vol. 7, No. 4: 2015, ISSN 1916-9639, E-ISSN 1916-9647**

## Confirmation of the electromagnetic gravity effect

Another physicist, Libor Neumann from Prague, performed a similar experiment and got the similar results. He published his findings in Physics Essays in 2017 in an article titled "Experimental verification of electromagnetic-gravity effect: Weight light and heat". He used a similar experiment that resembled more of a Cavendish pendulum experiment. Both ends of the pendulum were positioned beside a tub filled with light. The light in the empty tube produced a gravity effect as if the tube was filled with matter. The pendulum moved toward the tubes as if there was gravity created from the light. His experiment has yielded some additional info, and he found that each watt of power produces a 15 percent increase in gravity effect from the light. His confirmation of the light-gravity effect validates Louis Rancourt's experiments.

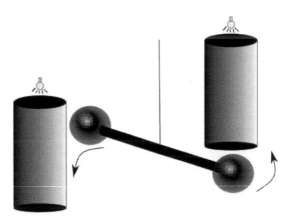

Figure 60

*Neumann, Libor. (2017). Experimental verification of electromagnetic-gravity effect: Weighing light and heat. Physics Essays. 30. 138-147. 10.4006/0836-1398-30.2.138.

Libor Neumann, in his paper, proposed an alternative explanation for this effect. He proposed that light created the gravity effect. This view contradicted Louis Rancourt's initial views but agrees with the current gravitational model. One subtle difference between Rancourt and Neumann's experiments is the time it took for the effects to manifest. This small difference holds a vital clue. In Rancourt's experiment, the laser was stopped, but the effect of the laser was still present for some time afterwards. If the electromagnetic waves had caused a conventional gravity effect, then the effect should have stopped immediately after the light was turned off. The weight should have returned to normal almost instantly. If the source of gravity has stopped, so should the effect.

In Rancourt's experiment, the effect lingered for some time after the laser was turned off. If the laser was interfering with the mechanism that produces the mass of the object, then this indicates the gravitational mass of the object is produced by an external mechanism. This does not agree with current physics models. Rancourt's experiment and his initial conclusion played a pivotal role in the development of the pushing field theory. It introduced me to LeSage's gravitational model and guided me to an entirely new outlook on physics.

# EXPERIMENTS TO RECONSIDER

This brings us back to the famous Michelson-Morley experiment that did not detect an interference pattern due to the earth's movement through the Aether. The Aether was used in the 1800s to explain the propagation of light, similar to how sound propagates through air. The idea behind the Aether in some ways is similar to the pushing field, but there is a fundamental difference between the two. Although the pushing field shares a similar idea, it differs in that the field is not a medium like a liquid but a dynamic field that interacts with matter. This interaction produces gravity as well as gravitational time dilation. As the waves of the pushing field propagate through space, they are restricted to the speed of light. This restriction of the field can be understood as the fundamental cause of time. This restriction prevents the waves in the pushing field from having an infinite velocity by acting as a "drag" on the waves. The restriction can be seen as another component of the pushing field as time; one component is force, and one component is time.

Mass has an effect on the local pushing field, similar to how a charged particle has an effect locally on the electric field. This effect causes a local disruption in the pushing field emanating from the mass. In addition, the idea of an infinite number of frequencies in the pushing field oscillations removes the possibility of measuring an absolute reference.

The Michelson-Morley experiment was designed to measure a fluidic Aether, not a fluctuating field. In that respect, the experiment did accomplish its task and did not detect the existence of the Aether. The experiment should not be used to measure a relative velocity in reference to a pushing field for a number of reasons. An experiment that is more indicative of the pushing field is the Sagnac experiment and the Foucault pendulum.

## Sagnac Experiment

An experiment that I believe agrees with the pushing field theory is the Sagnac experiment, which was first performed in 1911. In brief, the experiment used the interference of a split pulse of light, which traveled down different paths in the opposite direction of a light detector. The device, when rotating on its axis, will cause one pulse to arrive before the other pulse at the detector.

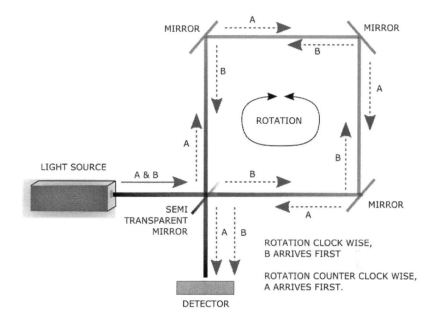

Figure 61

*(The figure above demonstrates the measuring apparatus of the Sagnac experiment)*

This experiment demonstrates the Sagnac effect. Similar devices are widely used as inertial navigation systems in planes and submarines. This experiment demonstrates how photons' velocities are relative to each other and not a measuring device. Rotating the measuring device does not affect the photons' velocities, but it does affect the arrival time, and interference pattern.

## Foucault pendulum

In 1851, Foucault's experiment demonstrated the rotation of the earth by swinging a pendulum back and forth over a reference on the ground. The pendulum would swing on its own in a circle over a period of time. If this pendulum was positioned directly over the North Pole, it would swing back and forth and complete a full turn on its axis in one day. The pendulum does not turn at all, but, in fact, it is the earth that rotates on its axis. The pendulum swings independently and does not follow the earth's rotation.

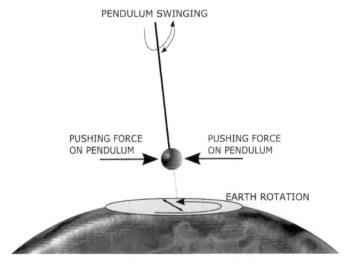

PENDULUM SWINGING

PUSHING FORCE
ON PENDULUM

PUSHING FORCE
ON PENDULUM

EARTH ROTATION

Figure 62

*(The Foucault pendulum does not follow the rotation of the earth on its axis.)*

I believe this indicates a mechanism is keeping the pendulum in its relative position. The only link between the pendulum and the earth is gravity and the tether that supports the swinging mass. It could be argued that it is inertia that keeps the pendulum from rotating with the earth. What mechanism communicates the information to the pendulum? In other words, how does the pendulum know not to rotate at the velocity of the earth? It is not gravity. If it was, how would the pendulum know not to rotate at the speed of the earth? Could it be the Higgs field? If it is, then an argument similar to the Aether could be made, as the Higgs field is static and not dynamic like the pushing field. If all reference points were removed (the stars, sun, and other planets), how could we know the earth is rotating and not the pendulum? The answer is in the motion of the pendulum. If the pendulum swings while rotating, it would move in a figure eight motion. That is not the behavior of Foucault's pendulum located at the Pantheon in Paris. This indicates that the universe has a relative "field" from which inertia is produced, and the pendulum follows that motion. The pendulum resist the change because of this field. There is an independent mechanism at work that exists in the universe that gives relative positions for the mechanism of inertia. I believe the Foucault pendulum is one more argument for the existence of the pushing field.

# CONCLUSION

My original intent for this project was to investigate alternative physics. As an amateur, I was curious to learn about forgotten physics and explore new concepts. This project took an entirely new direction when the evidence pointed me to an entirely new theory of physics. This should have never had happened unless something was indeed there to be found. Every time I investigate different aspects of the theory, I find more evidence supporting it. It is surprising that an alternative explanation for fundamental concepts exist and mirror somewhat the leading theories of today. This theory may in fact represent the unification theory long sought by physicists. Moreover, it is astonishing that theories such as the Le Sage and pilot wave theories support this new understanding. When I embarked on this journey of exploratory physics, I did not expect this outcome; I did not anticipate finding a rational explanation for phenomena that have for long been mysterious and illusive. I find it astounding that the universe starts with a very simplistic mechanism and expands in greater complexity as one effect builds on top of another. I am in awe at finding an answer to why the speed of light is so predominant in the universe and now see the inner workings of the universe as similar to a grandfather clock that ticks away as events unfold in a classical manner rather than a mysterious universe dependent on our observations. This ticking happens at the speed at which the pushing field acts on matter and massless particles in the universe; it is the common denominator in the universe. The pushing field is working behind the scenes as a clock, producing the mechanism gravity, inertia, and time. How else can we explain without a common phenomenon that underlies everything?

I will admit that this is a far-reaching idea, but is this not how such grand ideas start? This could be the next logical step in the evolution of physics. If we want to continue down this path, we have to be prepared that at some point, Einstein's relativity theory will be surpassed by another, just as his theory of relativity was an improvement over Newton's theory. They were both pioneers, blazing the way through the unknown. We have great admiration for these men and their work, and this theory encompasses their efforts.

Due to my ignorance, I initially made wrong assumptions. This provided me with a different interpretation of physics that led me down this path. I let the pieces of the puzzle fall where they seemed to belong. An entirely new view of physics emerged. There is many more avenues left to explore. This book is just the beginning, and follows in the footsteps of giants. It should demonstrates there is an entirely new understanding waiting to be discovered. One can only imagine the technological advancements that could follow from this unprecedented endeavor.

pushingfieldtheory@gmail.com

# REFERENCES

Roberg, Clement. M. (2018). A New perspective on gravity, inertial forces and time. Clement Roberge Publishing.

Roberge, Clement. M. (2017) Pushing Field Theory, ISBN 978-1-7750137-0-9

Elert, Glenn. (2018). Gravitational Potential Energy, The Physic Hypertextbook, https://physics.info/gravitation-energy/

Wikipedia contributors. (2018, November 29). Inertia. In *Wikipedia, The Free*

*Encyclopedia*. Retrieved 13:46, December 24, 2018,

from https://en.wikipedia.org/w/index.php?title=Inertia&oldid=871

210903

Wikipedia contributors. (2018, October 29). Newton's laws of motion.

In *Wikipedia, The Free Encyclopedia*. Retrieved 13:47, December 24,

2018,

from https://en.wikipedia.org/w/index.php?title=Newton%27s_laws

_of_motion&oldid=866282675

Wikipedia contributors. (2019, March 4). Special relativity. In *Wikipedia, The Free Encyclopedia*. Retrieved 10:44, March 17, 2019,
from https://en.wikipedia.org/w/index.php?title=Special_relativity&oldid=88
6100522

Wikipedia contributors. (2019, March 11). Luminiferous aether. In *Wikipedia, The Free Encyclopedia*. Retrieved 10:45, March 17, 2019,
from https://en.wikipedia.org/w/index.php?title=Luminiferous_aether&oldid
=887325048

Wikipedia contributors. (2019, March 8). Michelson–Morley experiment. In *Wikipedia, The Free Encyclopedia*. Retrieved 10:46, March 17, 2019,
from https://en.wikipedia.org/w/index.php?title=Michelson%E2%80%93Mo
rley_experiment&oldid=886807191

Georges-Louis Le Sage. (2017, July 16). In Wikipedia, The Free Encyclopedia. Retrieved 16:23, November 9, 2017, from https://en.wikipedia.org/w/index.php?title=Georges-Louis_Le_Sage&oldid=790834073


Wikipedia contributors. (2019, February 7). Lorentz transformation. In *Wikipedia, The Free Encyclopedia*. Retrieved 10:58, February 17, 2019, from https://en.wikipedia.org/w/index.php?title=Lorentz_transformation&oldid=882169716


Wikipedia contributors. (2019, March 13). General refinal

vity. In *Wikipedia, The Free Encyclopedia*. Retrieved 10:51, March 17, 2019, from https://en.wikipedia.org/w/index.php?title=General_relativity&oldid=887622704


Wikipedia contributors. (2018, November 13). Ives–Stilwell experiment. In *Wikipedia, The Free Encyclopedia*. Retrieved 13:37, February 17, 2019, from https://en.wikipedia.org/w/index.php?title=Ives%E2%80%93Stilwell_experiment&oldid=868668589


Wikipedia contributors. (2018, November 4). Sagnac effect. In *Wikipedia, The Free Encyclopedia*. Retrieved 16:08, February 23, 2019, from https://en.wikipedia.org/w/index.php?title=Sagnac_effect&oldid=867223322


Wikipedia contributors. (2019, March 8). Hafele–Keating experiment. In *Wikipedia, The Free Encyclopedia*. Retrieved 13:31, March 10, 2019, from https://en.wikipedia.org/w/index.php?title=Hafele%E2%80%93Keating_experiment&oldid=886756164

Rancourt, L. (2011). Effect of light on gravitation attraction. Physics Essays,

24(4), 557-561.

Rancourt, L. (2015) Further Experiments Demonstrating the Effect of Light on Gravitation. Applied Physics Research; Vol. 7, No. 4, ISSN 1916-9639 E-ISSN 1916-9647

Wikipedia contributors. (2019, January 9). Refractive index. In *Wikipedia, The Free Encyclopedia*. Retrieved 14:23, January 13, 2019, from https://en.wikipedia.org/w/index.php?title=Refractive_index&oldid=877543774

Wikipedia contributors. (2018, December 28). List of physical properties of glass. In *Wikipedia, The Free Encyclopedia*. Retrieved 16:18, January 27, 2019, from https://en.wikipedia.org/w/index.php?title=List_of_physical_properties_of_glass&oldid=875705180

Wikipedia contributors. (2019, February 1). Mercury (planet). In *Wikipedia, The Free Encyclopedia*. Retrieved 12:46, February 2, 2019, from https://en.wikipedia.org/w/index.php?title=Mercury_(planet)&oldid=881324053

Wikipedia contributors. (2019, January 20). Atomic orbital. In *Wikipedia, The Free Encyclopedia*. Retrieved 12:48, February 2, 2019, from https://en.wikipedia.org/w/index.php?title=Atomic_orbital&oldid=879287556

Aether (classical element). (2017, October 22). In *Wikipedia, The Free Encyclopedia*. Retrieved 16:22, November 9, 2017, from https://en.wikipedia.org/w/index.php?title=Aether_(classical_element)&oldid=806432153

Wikipedia contributors. (2019, June 21). James Clerk Maxwell. In *Wikipedia, The Free Encyclopedia*. Retrieved 11:45, June 23, 2019, from https://en.wikipedia.org/w/index.php?title=James_Clerk_Maxwell&oldid=902877569

Lorentz ether theory. (2017, October 5). In *Wikipedia, The Free Encyclopedia*. Retrieved 16:39, November 9, 2017, from https://en.wikipedia.org/w/index.php?title=Lorentz_ether_theory&oldid=803949200

Neumann, Libor. (2017). Experimental verification of electromagnetic-gravity effect: Weighing light and heat. Physics Essays. 30. 138-147. 10.4006/0836-1398-30.2.138.

Standard Model. (2017, November 28). In *Wikipedia, The Free Encyclopedia*. Retrieved 12:39, December 1, 2017, from https://en.wikipedia.org/w/index.php?title=Standard_Model&oldid=8125 65023

Photon. (2017, November 27). In *Wikipedia, The Free Encyclopedia*. Retrieved 13:10, December 1, 2017, from https://en.wikipedia.org/w/index.php?title=Photon&oldid=812344594

Beta decay. (2017, November 25). In *Wikipedia, The Free Encyclopedia*. Retrieved 14:21, December 1, 2017, from https://en.wikipedia.org/w/index.php?title=Beta_decay&oldid=81196542 6

Wave function. (2017, November 13). In *Wikipedia, The Free Encyclopedia*. Retrieved 14:29, December 1, 2017, from https://en.wikipedia.org/w/index.php?title=Wave_function&oldid=81007 0836

Probability density function. (2017, November 26). In *Wikipedia, The Free Encyclopedia*. Retrieved 14:32, December 1, 2017, from https://en.wikipedia.org/w/index.php?title=Probability_density_function &oldid=812136896

Quantum field theory. (2017, December 5). In *Wikipedia, The Free Encyclopedia*. Retrieved 21:25, December 17, 2017, from https://en.wikipedia.org/w/index.php?title=Quantum_field_theory& oldid=813807367

Spacetime. (2017, December 23). In *Wikipedia, The Free Encyclopedia*. Retrieved 13:19, December 29, 2017, from https://en.wikipedia.org/w/index.php?title=Spacetime&oldid=81675 2198

Polarizer. (2017, December 2). In *Wikipedia, The Free Encyclopedia*. Retrieved 12:50, February 4, 2018, from https://en.wikipedia.org/w/index.php?title=Polarizer&oldid=813167 249

Pilot wave. (2017, December 18). In *Wikipedia, The Free Encyclopedia*. Retrieved 22:42, February 4, 2018, from https://en.wikipedia.org/w/index.php?title=Pilot_wave&oldid=8160 18766

De Broglie–Bohm theory. (2018, January 26). In *Wikipedia, The Free Encyclopedia*. Retrieved 14:41, February 10, 2018, from https://en.wikipedia.org/w/index.php?title=De_Broglie%E2%80%9 3Bohm_theory&oldid=822530859

Wikipedia contributors. (2018, March 20). Foucault pendulum. In *Wikipedia, The Free Encyclopedia*. Retrieved 21:20, March 23, 2018, from https://en.wikipedia.org/w/index.php?title=Foucault_pendulum&ol did=831400254

Wikipedia contributors. (2018, March 22). Centrifugal force. In *Wikipedia, The Free Encyclopedia*. Retrieved 14:45, April 16, 2018, from https://en.wikipedia.org/w/index.php?title=Centrifugal_force&oldid=83 1927579

Wikipedia contributors. (2019, June 1). Quantum fluctuation. In *Wikipedia, The Free Encyclopedia*. Retrieved 10:08, June 8, 2019, from https://en.wikipedia.org/w/index.php?title=Quantum_fluctuation&oldid =899856572

Wikipedia contributors. (2019, August 6). Centripetal force. In *Wikipedia, The Free Encyclopedia*. Retrieved 11:08, August 17, 2019, from https://en.wikipedia.org/w/index.php?title=Centripetal_force&oldid =909535820

Wikipedia contributors. (2019, July 1). Le Sage's theory of gravitation. In *Wikipedia, The Free Encyclopedia*. Retrieved 11:11, August 17, 2019, from https://en.wikipedia.org/w/index.php?title=Le_Sage%27s_theory_of_gravitation&oldid=904295901

Wikipedia contributors. (2019, July 27). Gyroscope. In *Wikipedia, The Free Encyclopedia*. Retrieved 11:13, August 17, 2019, from https://en.wikipedia.org/w/index.php?title=Gyroscope&oldid=9080 69670

Wikipedia contributors. (2019, August 6). Precession. In *Wikipedia, The Free Encyclopedia*. Retrieved 11:36, August 17, 2019, from https://en.wikipedia.org/w/index.php?title=Precession&oldid=9095 96543

Wikipedia contributors. (2019, August 14). Quantum mechanics. In *Wikipedia, The Free Encyclopedia*. Retrieved 11:37, August 17, 2019, from https://en.wikipedia.org/w/index.php?title=Quantum_mechanics&ol did=910815189

Wikipedia contributors. (2019, August 11). Wave. In *Wikipedia, The Free Encyclopedia*. Retrieved 11:38, August 17, 2019, from https://en.wikipedia.org/w/index.php?title=Wave&oldid=91036177 5

Wikipedia contributors. (2019, July 26). Casimir effect. In *Wikipedia, The Free Encyclopedia*. Retrieved 11:40, August 17, 2019, from https://en.wikipedia.org/w/index.php?title=Casimir_effect&oldid=9 07978714

Wikipedia contributors. (2019, July 28). Van der Waals force. In *Wikipedia, The Free Encyclopedia*. Retrieved 11:40, August 17, 2019, from https://en.wikipedia.org/w/index.php?title=Van_der_Waals_force& oldid=908312008

Wikipedia contributors. (2019, July 30). Strong interaction. In *Wikipedia, The Free Encyclopedia*. Retrieved 11:41, August 17, 2019, from https://en.wikipedia.org/w/index.php?title=Strong_interaction&oldi d=908500285

Wikipedia contributors. (2019, August 16). Electron. In *Wikipedia, The Free Encyclopedia*. Retrieved 11:43, August 17, 2019, from https://en.wikipedia.org/w/index.php?title=Electron&oldid=911037 936

Wikipedia contributors. (2019, August 14). Magnetic field. In *Wikipedia, The Free Encyclopedia*. Retrieved 11:43, August 17, 2019, from https://en.wikipedia.org/w/index.php?title=Magnetic_field&oldid=9 10838235

Wikipedia contributors. (2019, July 12). Electric field. In *Wikipedia, The Free Encyclopedia*. Retrieved 11:44, August 17, 2019, from https://en.wikipedia.org/w/index.php?title=Electric_field&oldid=905933972

Wikipedia contributors. (2019, August 15). Maxwell's equations. In *Wikipedia, The Free Encyclopedia*. Retrieved 11:45, August 17, 2019, from https://en.wikipedia.org/w/index.php?title=Maxwell%27s_equations&oldid=910958352

Wikipedia contributors. (2019, July 4). Classical electromagnetism. In *Wikipedia, The Free Encyclopedia*. Retrieved 11:45, August 17, 2019, from https://en.wikipedia.org/w/index.php?title=Classical_electromagnetism&oldid=904838939

Wikipedia contributors. (2019, August 17). Electromagnetic radiation. In *Wikipedia, The Free Encyclopedia*. Retrieved 11:46, August 17, 2019, from https://en.wikipedia.org/w/index.php?title=Electromagnetic_radiation&oldid=911210924

Wikipedia contributors. (2019, August 11). Wavelength. In *Wikipedia, The Free Encyclopedia*. Retrieved 11:47, August 17, 2019, from https://en.wikipedia.org/w/index.php?title=Wavelength&oldid=910360930

Wikipedia contributors. (2019, August 7). Speed of light. In *Wikipedia, The Free Encyclopedia*. Retrieved 11:49, August 17, 2019, from https://en.wikipedia.org/w/index.php?title=Speed_of_light&oldid=909811814

Wikipedia contributors. (2019, August 17). Transparency and translucency. In *Wikipedia, The Free Encyclopedia*. Retrieved 11:49, August 17, 2019, from https://en.wikipedia.org/w/index.php?title=Transparency_and_translucency&oldid=911200745

Wikipedia contributors. (2019, August 16). Atom. In *Wikipedia, The Free Encyclopedia*. Retrieved 11:50, August 17, 2019, from https://en.wikipedia.org/w/index.php?title=Atom&oldid=911055558

Wikipedia contributors. (2019, August 17). Refractive index. In *Wikipedia, The Free Encyclopedia*. Retrieved 11:53, August 17, 2019, from https://en.wikipedia.org/w/index.php?title=Refractive_index&oldid=911187604

Wikipedia contributors. (2019, July 26). Tests of general relativity. In *Wikipedia, The Free Encyclopedia*. Retrieved 11:53, August 17, 2019, from https://en.wikipedia.org/w/index.php?title=Tests_of_general_relativity&oldid=907968866

THANK YOU


THE END

Printed in Great Britain
by Amazon

43444142R10059